SpringerBriefs in Operations Management

Series Editor

Suresh P. Sethi
The University of Texas at Dallas, TX, USA

More information about this series at http://www.springer.com/series/13082

Jagdeep Singh • Harwinder Singh

Strategic Implementation of Continuous Improvement Approach

Improving the Performance of Small and Medium-Sized Enterprises

Springer

Jagdeep Singh
Industrial and Production Engineering
Department
Dr B R Ambedkar NIT
Jalandhar, Punjab, India

Harwinder Singh
Mechanical Engineering Department
Guru Nanak Dev Engineering College
Ludhiana, Punjab, India

ISSN 2365-8320 ISSN 2365-8339 (electronic)
SpringerBriefs in Operations Management
ISBN 978-3-319-93120-3 ISBN 978-3-319-93121-0 (eBook)
https://doi.org/10.1007/978-3-319-93121-0

Library of Congress Control Number: 2018944138

This Springer imprint is published by the registered company Springer International Publishing AG part of Springer Nature.
The registered company address is: Gewerbestrasse 11, 6330 Cham, Switzerland

Preface

In this competitive world, industrial organizations are focusing to enhance their productivity through system simplification, organization potential, and incremental improvements by using modern techniques like continuous improvement (CI) or KAIZEN. The present book seems to fulfill an obvious need to write for a variety of reasons. Firstly, although the concept of KAIZEN is familiar to engineers, their treatment of it has always been narrowly focused. Secondly, many people, including engineers, are still wrestling with the difficulties associated with the concept, its applications, and implementation on the shop floors. Thirdly, the role of engineers seems to be more and more positively enhanced nowadays. Their involvement is much more widely spread than previously thought. As KAIZEN tends to focus on business organizations as total dynamic systems, its portrayal in the present form should help engineers appreciate the level of penetration and wide implications. KAIZEN is an evolutionary but umbrella concept that covers all aspects of business units.

KAIZEN is a means to drive the business rather than just being a loose input. It is unfortunate that cynics who consider CI to be just a craze have failed to visualize the evolutionary process by which considerable growth in possible.

This book aims to present a comprehensive approach which is intended to broaden the knowledge of engineers and others concerned with KAIZEN and its implications on their working environment. It is also intended for students and researchers who have to wrestle with this concept in the course of their various studies.

Nonetheless, the authors are grateful to all the people whose ideas have helped shed some light on specific areas of KAIZEN and whose contributions in this field are valued and greatly appreciated by all those who aspire to progress and advancement.

The authors are also grateful to their parents for their support and understanding at all stages of writing this book. Finally, the authors devote this work to THE ALMIGHTY GOD, the only one who has granted them the willingness and ability to successfully complete this book.

Jalandhar, Punjab, India — Jagdeep Singh
Ludhiana, Punjab, India — Harwinder Singh

Contents

About the Authors

Jagdeep Singh holds a bachelor's degree in mechanical engineering, master's degree in production engineering, and Ph.D. in mechanical engineering. The author has published a good number of research papers in international and national journals. The author is working as an assistant professor in the Industrial and Production Engineering Department at Dr B.R. Ambedkar NIT Jalandhar, Punjab. The field of interest of the author is management of production systems and operations management.

Harwinder Singh holds a bachelor's degree in mechanical engineering, master's degree in production engineering, master's degree in business administration, and Ph.D. in mechanical engineering. The author has published a good number of research papers in international and national journals. The author is working as a professor in the Department of Mechanical Engineering at Guru Nanak Dev Engineering College, Ludhiana, Punjab. The field of interest of the author is management of production systems, operations management, and decision making.

Abstract

Continuous improvement (CI) is a technique to improve the performance at every level of operation, in every functional area of an organization, by maximum utilization of available resources. It is a management approach that focus on enhancement of manufacturing processes through incremental changes. It is a management approach of identifying and eliminating waste (all non-value-added activities) through incremental improvements by reducing the time associated with setup in pursuit of perfection. CI programs have evolved from traditional manufacturing focused systems that concentrate on the production line to reduce waste and improve the product quality, into comprehensive, systematic methodologies that is focused on the entire organization. To achieve excellence in every sphere of the world is not a fortnightly process. If it is not taken care of, the competencies gained by the organization might fade away with the passage of time. The organization must consistently send a strong message to employees that continuous improvement strategies are the core values of the company. Different strategies of CI approach are being used by the manufacturing organizations to improve the performance of current manufacturing system processes.

The present study is an attempt to check the performance of KAIZEN technique in small–medium enterprises (SMEs) of Northern India. A survey of 101 SMEs has been performed in Northern India. The level of usage, level of importance of different strategies, level of importance of different enablers, and the important benefits achieved from KAIZEN implementation have been identified. Correlation analysis shows significant relationship between different CI strategies and benefits of CI approach. SMEs use total quality management (TQM) strategies to its highest maturity level. Two case studies have been performed to ascertain the important benefits occurred after CI strategic implementation. CI strategies including value stream analysis, redesign, suggestion system, failure mode effect analysis, process flow mapping, recognition, minor stoppage elimination, and customer quality, cost and delivery analysis (QCD) through holistic CI implementation program can significantly contribute in harnessing incremental improvements in the organization, thereby providing a sound platform for the organization to compete effectively in the dynamic environment. CI activities through strategic implementation can contribute towards quality enhancements by eliminating deteriorations in production systems but also have negative impact on the safety-related issues of employees. The study critically reveals the contributions of CI strategic implementation for achieving manufacturing performance improvement through quality initiatives in the SMEs of Northern India. The study reported in this work has revealed that there is a benefit of CI strategic implementation in the Indian industry and there is a need to develop an implementation plan to foster CI implementation practices and procedures. Overall equipment effectiveness (OEE) parameters seek to encourage the setting of ambitious, but attainable, realistic goals for raising the OEE by focusing on the losses related to availability, performance, and quality of a system. It prepares the plant to meet the challenges of competitive manufacturing by adoption and implementation of a well-conceived plan with the help of preventive maintenance teams for effective equipment maintenance.

Strategic implementation of CI approach systematically aims at improving the OEE using setup time reduction. Six Sigma approach is helpful in achieving process improvement and thereby achieving different goals of productivity and the organization. The successful implementation of single-minute exchange of die (SMED) and other CI strategies are the key to a competitive advantage for any manufacturer that produces, prepares, processes, or packages a variety of products. The reduction of defective part million opportunity (DPMO) level from 75655.25 to 455.26 has been obtained to enhance the performance of manufacturing unit. The net savings of 17.65 lacs per year have been obtained after the successful implementation of the DMAIC approach. Six Sigma provides an effective mechanism to focus on customer requirements, through improvement of process quality. CI has proven to be a means to supplement the concerted improvement efforts by addressing equipment and other related problems that affect the performance of the manufacturing system.

CI implementation in the enterprise has demonstrated the effective realization of optimized equipment effectiveness, minor stoppage elimination, and promotion of autonomous operator maintenance through day-to-day activities involving the total workforce. CI has helped the enterprise in improving the synergy between the production department and rest of the improvement functions, resulting in elimination of defects, improved process reliability like OEE, affecting cost reductions thereby strengthening sustainability efforts of the organization to meet cut-throat global competition for business excellence. It therefore becomes pertinent for the successful organizations to integrate improvement initiatives into their manufacturing strategy for realization of organizational objectives. The successful CI implementation program can facilitate the manufacturing organization's quest for achieving enhanced manufacturing performance leading to the firm's performance improvement. Thus, in today's competitive environment CI approach has proven to be an effective global strategy, for Indian organizations, rendering firms a consistent enhancement of performance in terms of achieving manufacturing performance improvement.

Implementation plan reveals that the implementation of CI strategies leads to the success of the organization by focusing on the losses related to availability, performance, and quality of a system. Structural equation modeling (SEM) shows dependency of one initiative on other initiative of CI approach. SEM is a powerful tool that enables researchers to go beyond factor analysis into the arena of determining whether one set of unobserved constructs is related to another set of constructs. Customer and supplier relations play a significant role in improving the performance of the firm and hence improve the manufacturing system processes showing high positive correlation. SEM investigates the impact of three sets of antecedent factors on the firm performance. Customer satisfaction regarding cost, delivery, and quality plays an role important role in improving the firm performance by obtaining different benefits including organization achievement, increased productivity, improved quality, reducing cost, improving safety, and timely delivery of the product manufactured.

Chapter 1
Introduction

1.1 Meaning and History of KAIZEN Approach

KAIZEN is a Japanese word that has become common in many Western companies; the word indicates a process of continuous, incremental improvement of the "standard" way of work (Chen, Dugger, & Hammer, 2000). The term came from Gemba KAIZEN in Japanese meaning "continuous improvement" (Yan-jiang, Dan, & Lang 2006). It is a compound word involving two concepts as shown in Fig. 1.1.

Originally a Buddhist term, KAIZEN came from the words, "Renew the heart and make it good." Therefore, adaptation of the KAIZEN concept also requires changes in "the heart of the business," corporate culture, and structure, since KAIZEN enables companies to translate the corporate vision in every aspect of a company's operational practice. KAIZEN is often translated in the West as ongoing continuous improvement (CI) (Malik & YeZhuang, 2006).

The roots of modern improvement programs can be traced back to initiatives undertaken in several companies in the 1800s, where management encouraged employee-driven improvements, and incentive programs were set in place to reward employees that brought about positive changes in the organization (Bhuiyan & Baghel, 2005). KAIZEN was originated from Japan in 1950 when management and government acknowledge that there is a problem in the current confrontational management system and a pending labor shortage. Japan sought to resolve this problem through cooperation with the workforce. The groundwork was established in the labor contracts championed by the government and is taken up by most major companies, which had introduced lifetime employment and guidelines for gainsharing distribution of benefits for the company development. This contract remains the background for all KAIZEN activities providing the necessary security to ensure confidence in the workforce (Brunet, 2000). Firstly, it was applied by Imai in 1986 to improve efficiency at Toyota Carmaker Company in response to increase in competition. Since then, KAIZEN had become a part of the Japanese manufacturing system processes and has contributed highly to the manufacturing success (Ashmore, 2001).

J. Singh, H. Singh, *Strategic Implementation of Continuous Improvement Approach*, SpringerBriefs in Operations Management,
https://doi.org/10.1007/978-3-319-93121-0_1

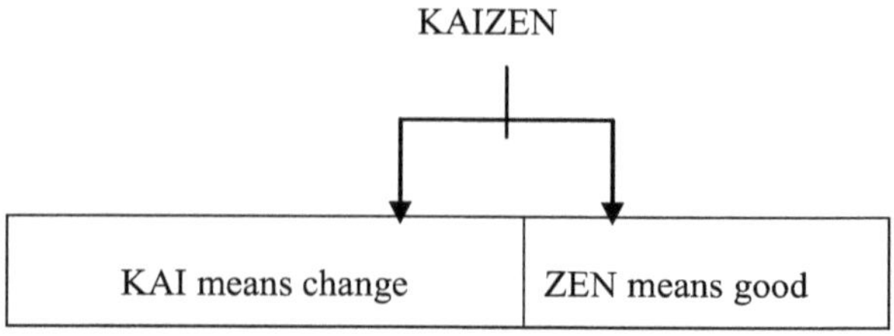

Fig. 1.1 Meaning of KAIZEN (source: Newitt, 1996)

As Indicated by Kaye and Anderson (1999), the approaches to quality have evolved significantly since the beginning of this century. Four major quality eras have been identified, viz. inspection, statistical quality control, quality assurance, and strategic quality management. Each quality era was built on the previous era. During the first three eras quality was viewed as a problem to be solved and the focus was on the internal operations of the organization. In the 1980s, quality was seen for the first time as a competitive opportunity, a strategic weapon that could be used against competitors. Thus, in the fourth quality era the focus was on the customer and the organization was more proactive in anticipating and responding to both customer and market needs. Continued top management commitment and involvement in quality was seen as paramount to ensure that quality was fully integrated into business strategy and plans and that it was adequately deployed throughout the organization. The strategic quality management approach, however, is still inadequate to meet today's rapidly changing business environment, characterized by uncertainty and unpredictability. To meet these challenges, a fifth quality era—competitive continuous improvement—has been identified.

1.2 KAIZEN and Innovation

There are two types of focus on process that are opposing: one of them is the gradual improvement and the other is innovation; the Japanese enterprises favor the gradual improvement and the Western enterprises favor the large jumps with invention. In spite of KAIZEN not demanding great investment for its implementation, it demands continuous efforts and commitment. In ideal conditions, KAIZEN can be compared to a ramp, since the progress is gradual, and the innovation to a step. Improvement can be broken down into KAIZEN and innovation (Hammer, Champy, & Tathan, 1993). KAIZEN signifies small improvements as a result of ongoing efforts. Innovation involves a drastic improvement as a result of a large investment of resources in new technology or equipment. In the context of KAIZEN, management has two major functions: maintenance and improvement. Maintenance refers to activities directed towards maintaining current technologies, managerial and operating standards, and upholding such standards through training and discipline. Under its maintenance function, management performs its assigned tasks so that everybody can follow standard operating procedure. Improvement, meanwhile, refers to activities directed towards elevating current standards as shown in Fig. 1.2.

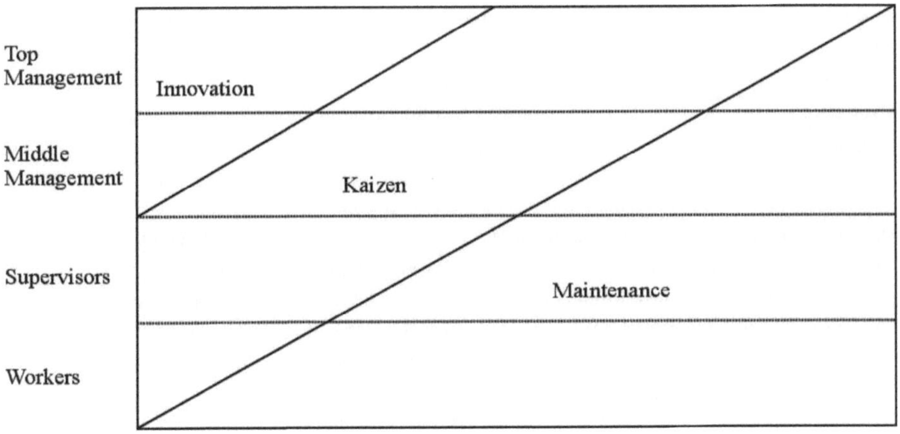

Fig. 1.2 Improvement broken down into KAIZEN and innovation (source: Imai, 1997)

In spite of KAIZEN demanding great investment for its implementation, it demands continuous efforts and commitment. In ideal conditions, KAIZEN can be compared to a ramp, since the progress is gradual, and the innovation to a step. There are two types of focus on progress that are opposing; one is the gradual improvement called KAIZEN and the other is large steps commonly known as innovation (Shingo, 1985). KAIZEN may facilitate innovation, but the innovation refers to incremental innovation or small-scale innovation conducted by blue-collar workers and not to disruptive innovation such as value innovation (Setijono, 2010). Table 1.1 presents the comparison of KAIZEN and innovation.

However, the KAIZEN strategy does call for continuous efforts and commitment from all levels of management. Thus KAIZEN calls for a substantial management commitment of time and effort. Innovation, which is implemented by abrupt changes, and KAIZEN, which is a gradual process, are alternative means of introducing improvement. KAIZEN means small improvements made as a result of continuing effort. Innovation involves a drastic improvement made as a result of large improvements in new equipment or technology (Wittenberg, 1994).

1.3 SMEs in Indian Context

In India, 95% of industrial units are in the small–medium enterprise (SME) sector with 40% value addition in the manufacturing sector (Singh & Khanduja, 2010). As suggested by Deshmukh and Chavan (2012):

- A small enterprise is an enterprise where the investment in plant and machinery (original cost excluding land and building and the items specified by the Ministry of Small Scale Industries) is more than Rs. 25 lakhs but does not exceed Rs. 5 crores

Table 1.1 Comparison of KAIZEN and innovation (source: Karkoszka & Szewieczek, 2007; Radharamanan, Godoy, & Watanabe, 1996)

KAIZEN	Innovation
Creativity	Ability to adaptation
Individualism	Teamwork
Orientation towards the specialists	Orientation towards person having no specialized qualifications
Attaching great value to the general matters	Attaching great value to the details
Orientation towards techniques	Orientation towards human
Information directed to the chosen persons	Free-for-all information, generalized
Orientation to the individual section	Interdepartmental orientation
Searching for the new technology	Basing on the existing technologies
Limited feedback	Strong feedback
Short-time effect	Long-time effect
Participation of several chosen "leaders"	Participation of every worker
Adaptation to the fast-growth-rate economy	Adaptation to the slow-growth-rate economy
Big investment needed	Small investment needed
Results in the aspect of profit as the estimation criterion	Processing estimation criterion
Continuous and incremental	Intermittent and not incremental
Collective team efforts, system focus	Strong individuality, individual ideas, and efforts
Maintenance and improvement	Rejection and rework
Know-how and conventional updating	Technological advances, new inventions, and new theories
Persons	Technology
Demand less investment; however, greater effort to maintain	Demand large investment; however, less effort to maintain

- A medium enterprise is an enterprise where the investment in plant and machinery (original cost excluding land and building and the items specified by the Ministry of Small Scale Industries) is more than Rs. 5 crores but does not exceed Rs. 10 crores

Never before, the role of small–medium enterprises (SMEs) was as important as it is today. SMEs throughout the world have made remarkable impact on their respective local economies. Though the basic nature of business by these SMEs has remained more or less same, today's competitive business does not allow them to remain unimaginative. Most of the governments throughout the world have protected SMEs by delineating protectionist policies; however, the globalization and free-trade initiatives have diluted the advantage SMEs enjoyed. The primary reason for the protectionist policies was their positive role in employment generation (safeguarding the social fabric of the particular geographical region), foreign exchange earnings, etc. However, today the rise in economic activities and competition across the globe demands that the products and services offered by SMEs should be of best quality (as these SMEs now compete with their global peers). The developments of world have shown great interest in the possible contributions of SMEs in the social and economic development. However, as in most industrialized countries, the SME sector are traditionally operating under significant constraints such as financial constraints, nonfinancial constraints, lack of skilled human resources, and constraints relating to the enabled environment. Since the SMEs often act as suppliers to a bigger company (which makes them an important link in the local as well as global supply chains), the production/assembly lines of the bigger companies today are more dependent on the SMEs (especially in view of the outsourcing of operations where the company does not have its core competency). Thus, the SMEs need to be more predictive regarding their ability to consistently supply different products to the bigger companies (Deshmukh & Lakhe, 2010). The Japanese concept of KAIZEN (continuous improvement) bridges scientific and humanistic management philosophies by focusing on areas such as recognition, autonomy, training, and development of individuals. This concept, already put into action by major manufacturers, can be implemented in SMEs to improve their overall operational performance. Integrating various functional groups within a manufacturing organization can also improve both productivity and quality (Gunasekaran, Korukonda, Virtanen, & Yli-Olli, 1994).

1.4 Definitions of Continuous Improvement Approach

Different researchers view continuous improvement approach from different angles; it carries different meaning to different researchers in different situations. The definitions suggested by different researchers have been discussed below:

Bhuiyan, Baghel, and Wilson (2006) defined, "Continuous Improvement does not mean only repeated small improvements. Continuous Improvement means repeated improvements of any size." Brunet and New (2003) defined that KAIZEN is the Japanese word for improvement, carrying the connotation in industry of all the uncontracted and partially contracted activities which take place in the Japanese workplace to enhance the operations and the environment. Caffyn (1999) defined that CI is a company-wide process of focused and continuous incremental innovation: in other words, mass involvement in making relatively small changes, which are directed, towards organizational goals on an ongoing basis. As defined by Doolen, June, Akan, Eileen, and Jennifer (2003) the Japanese meaning of KAIZEN is "to take apart and put back together in a better way." Continuous improvement can be defined as small incremental changes in productive processes or in working practices that permit an improvement in some indicator of performance (Marin-Garcia, Val, & Martin, 2008). CI is a collection of activities that constitute a process intended to achieve improvement. In manufacturing, these activities primarily involve simplification of production processes, chiefly through the elimination of waste (Jha, Noori, & Michela, 1996). Continuous improvement (CI) can be defined as "the planned, organized and systematic process of ongoing, incremental and company-wide change of existing practices aimed at improving company performance" (Jorgenson, Boer, & Gertsen, 2003). Yeo, Goh, and Xia (1995) defined KAIZEN as a management tool aimed at the reduction of defects through prevention. It is directed at motivating people to prevent mistakes by developing a constant, conscious desire to do their job right the first time. Kossoff (1993) asserted that "total quality is defined as the unrelenting pursuit of continuous improvement which is realized by accessing and utilizing the concerted knowledge and experience of managers and employees at all levels." Bessant and Caffyn (1994) defined the CI concept as "an organization–wide process of focused and sustained incremental innovation." Many tools and techniques are developed to support these processes of incremental innovation. The difficulty is the consistent application of continuous improvement philosophy and the CI tools and techniques. As an organization-wide process, CI requires the efforts of all employees on every level. According to Reid (2006), the philosophy of continuous improvement (CI) embraces the fundamental idea that whatever success an organization may have had in the past, in terms of customer satisfaction, it has not reached its final destination. Because the needs and wants of their customers are in a constant state of flux, organizations are never able to fully attain their ultimate goal.

1.5 Different CI Strategies

Implementation of CI approach involves different strategies so as to improve the performance of manufacturing system processes. These strategies eliminate waste where value is added and plays an important role in improving manufacturing system operations. These strategies are universal in nature and it involves all aspects of manufacturing. These include the generation of an appropriate improvement environment, which includes total commitment at all levels of management and within the workforce. Success stories reveal that it requires highly skillful efforts from the employees of the organization to fully implement the system. However, awareness among employees regarding the different techniques that are involved in implementing these strategies, various technical aspects behind these strategies, and utilization of these strategies in different circumstances plays an important role in implementing such incremental improvement strategies. Figure 1.3 presents a total system approach to continuous improvement or eight organizational strategies and sub-strategies of CI approach.

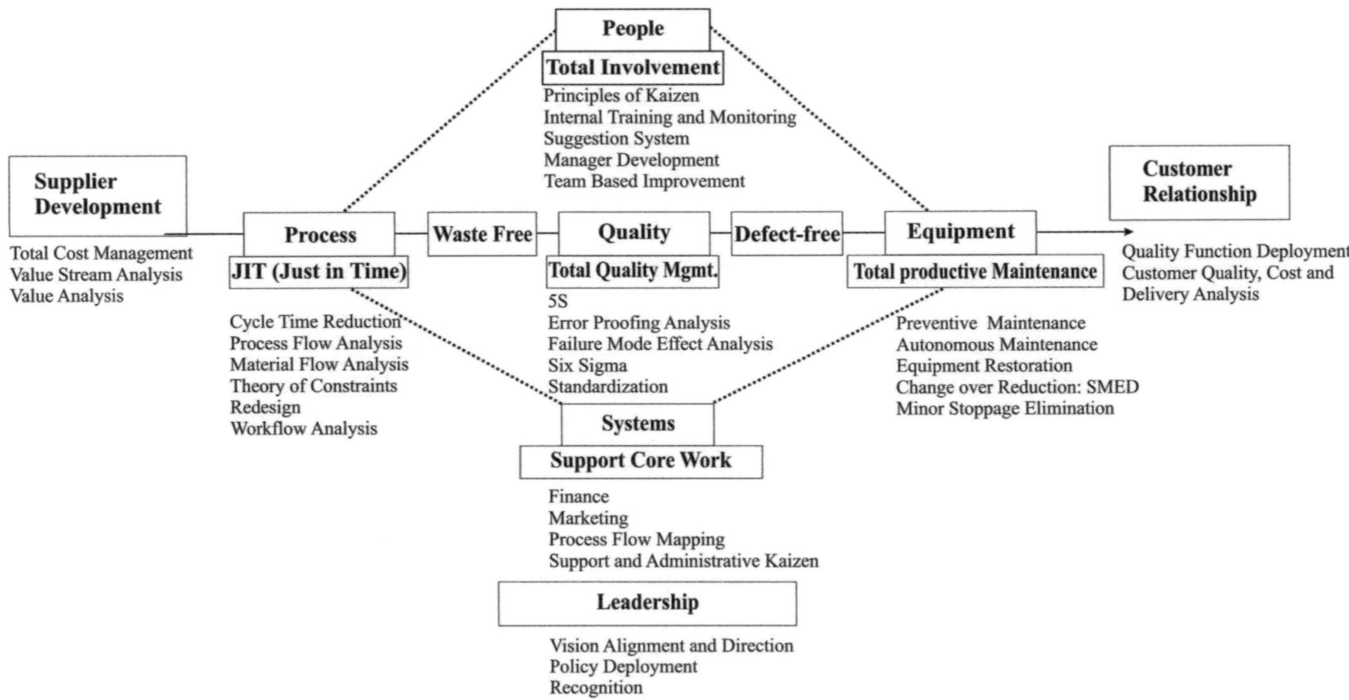

Fig. 1.3 Total system approach to continuous improvement (source: https://www.google.co.in/?gws_rd=cr&ei=e_RuU77lHMO0uASoyoHwCQ#q=8+organizational+elements+of+kaizen)

Chapter 2
Literature Review

2.1 Introduction

In today's highly dynamic and rapidly changing environment, the global competition among organizations has led to higher demands on the manufacturing organizations. The manufacturing industry has experienced an unprecedented degree of change in the last three decades, involving drastic changes in management approaches, product and process technologies, customer expectations, supplier attitudes, as well as competitive behavior. These challenges are forcing the manufacturing organizations globally to foster high reliability, availability, and maintainability in the manufacturing systems by implementation of various strategic and proactive market-driven strategies to remain competitive in a highly dynamic environment (Ahuja & Khamba, 2008). With increased competition, attention has been focused on improving efficiency of manufacturing operations to meeting market conditions in terms of delivery, cost, and quality (Yamashina, 1995). To meet the continuously changing environment, satisfy customer, and retain market share, incremental improvements of manufacturing system processes have become necessary (Shingeo, 1988). It includes collection of activities aiming at simplifying production processes through elimination of different types of waste (Terziovski, 2002). It is based on making incremental innovation supported by different types of practices (Malik, Li-bin, YeZhuang, & Xiao-lin, 2007). Companies are therefore applying continuous improvement practices to enhance their systems and operations.

2.2 Ongoing Continuous Improvement Process

KAIZEN is the Japanese word for improvement, carrying the connotation in industry of all the uncontracted and partially contracted activities which take place in the Japanese workplace to enhance the operations and the environment. KAIZEN

J. Singh, H. Singh, *Strategic Implementation of Continuous Improvement Approach*, SpringerBriefs in Operations Management,
https://doi.org/10.1007/978-3-319-93121-0_2

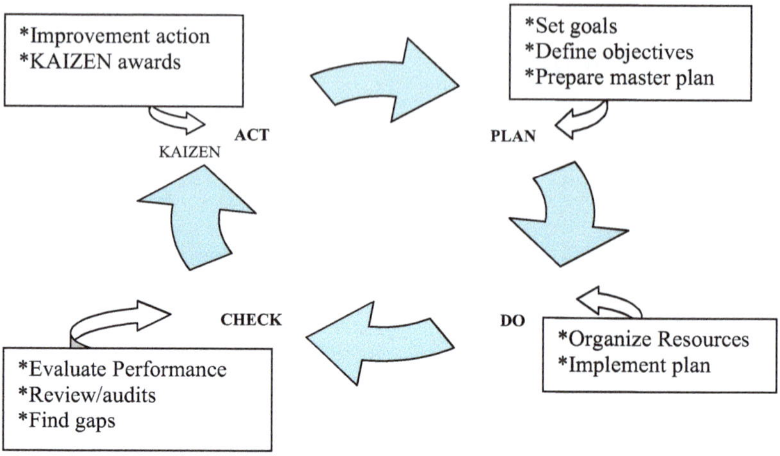

Fig. 2.1 PDCA cycle for ongoing improvement process (source: Grunberg, 2003)

epitomizes the mobilization of the workforce, providing the main channel for employees to contribute to their company's development. In isolation, the concept seems simple "with every pair of hands, you get a free brain." Improvement should be a way of corporate life. It is certainly central to many of the fashionable schools of management thinking such as JIT, TQM, and BPR. Change may be continuous or discontinuous. Improvement can be categorized as either small incremental change (KAIZEN) or innovative step change (process reengineering). The two are complementary, not mutually exclusive, routes to progress (Bond, 1999). The following concepts form the basis of the Deming's PDCA virtuous cycle of improvement:

- Plan—study current situation and develop changes for improvement.
- Do—pilot measures on a trial basis.
- Check—examine the effect of changes to see if the desired result is achieved.
- Action—standardize on a permanent basis as shown in Fig. 2.1.

Brunet and New (2003) emphasized different key features of ongoing CI process as shown below:

- That KAIZEN is continuous—which is used to signify both the embedded nature of the practice and its place in a never-ending journey towards quality and efficiency
- That it is usually incremental in nature, in contrast to major management-initiated reorganizations or technological innovation (e.g., the installation of new technology or machinery)
- That it is participative, entailing the involvement and intelligence of the workforce, generating intrinsic psychological and quality of work-life benefits for employees

CI process comprises four different phases: studying the current situation, collecting the data necessary to propose the suggestions for improvement; setting in motion trials of the proposals selected; checking whether the proposal tested is giving the expected results; and implementation and standardizing the proposal with the necessary modifications (Lawler, 1991).

2.3 PDCA and SDCA Cycles of Continuous Improvement

Watson (1986) described that KAIZEN strategy depends mainly on human efforts to improve results, and this requires process improvement. A process-oriented approach, referred to as the PDCA cycle, is used for process improvement. The origin of PDCA cycle or Deming cycle can be traced back to the eminent statistics expert Shewhart in the 1920s. Shewhart introduced the concept of plan, do, see. The late TQM guru Deming modified the Shewhart cycle as plan, do, study, act. The Deming cycle is a continuous quality improvement model consisting of a logical sequence of these four repetitive steps for continuous improvement and learning: plan, do, check, and act. The PDCA cycle is also known as Deming cycle, the Deming wheel of continuous improvement spiral. In plan phase, the objective is to plan for change, analyze, and predict the results. In do phase, the plan is executed by taking small steps in controlled circumstances. In study/ check phase the results are studied. Finally in act phase, the organization takes action to improve the process. These concepts form the basis of virtuous cycle of improvement:

- Plan—study the current situation and develop changes for improvement.
- Do—pilot measures on a trial basis.
- Check—examine the effect of changes to see if the desired result is achieved.
- Action—standardize on a permanent basis.

The main purpose of this cycle is to iron out abnormalities in the resulting work process and bring it back to harmony before moving to a new improving cycle. In other words, the standardizing cycle maintains current work processes, while the improving cycle improves them. The two cycles—PDCA and SDCA (standardize-do-check-act) cycles—revolve regularly to spread a culture of CI as a standard practice within an organization as shown in Fig. 2.2.

2.4 Principles of CI Approach

Customer-driven improvement is explicit to CI strategies, as all efforts are made to improve the performance of the product and process. It is an integrative philosophy of management for continuously improving the quality of products or services and processes to achieve customer satisfaction. Essentially, CI strategies are primarily concerned with every aspect of the company's activities and are based on the principles described in Table 2.1 (Love et al., 1998).

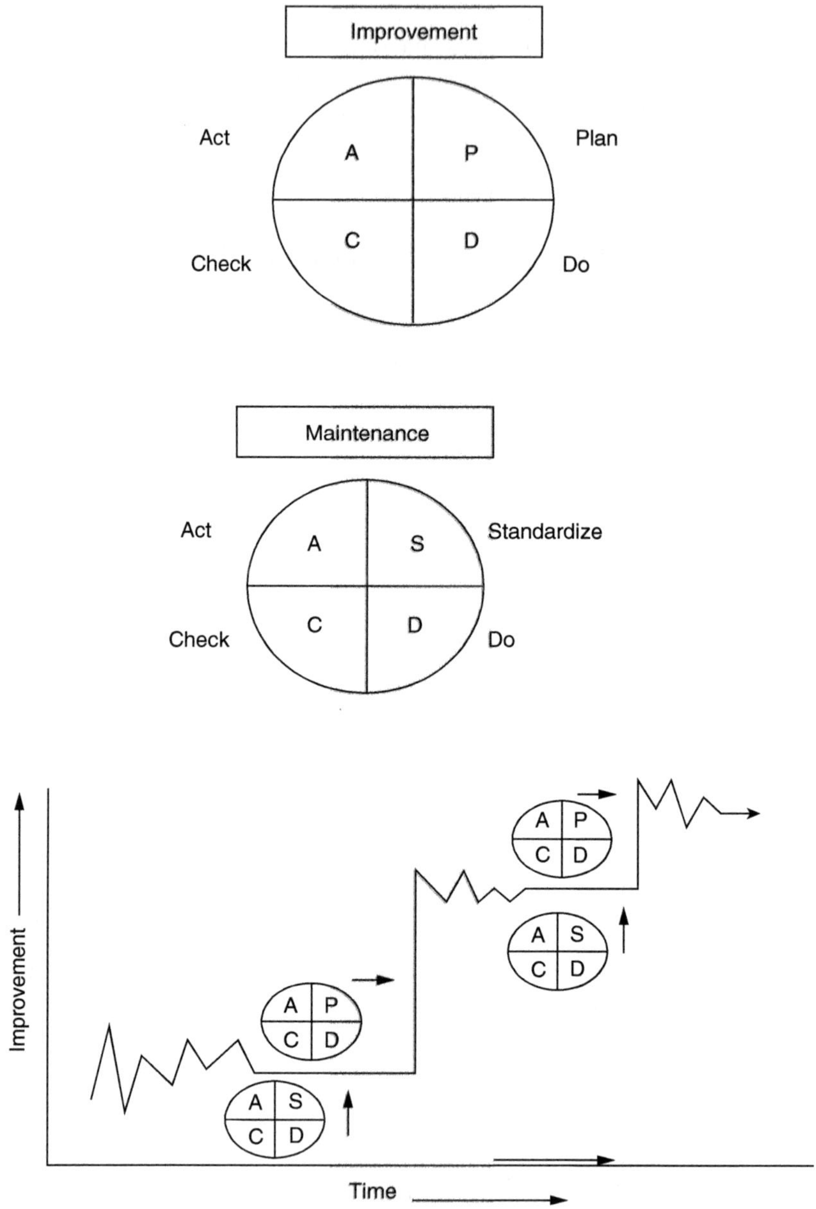

Fig. 2.2 PDCA and SDCA cycle (source: Locke & Jain, 1995)

Table 2.1 Principles of CI and their description

Principles	Description
Customer-driven organization	A company depends on its customers and therefore should understand current and future customer needs, meet customer requirements, and strive to exceed customer expectation
Leadership	Leaders establish unity of purpose, direction, and internal environment of organization. Leaders should try to fully involve people in achieving the organization's goals and objectives
People participation	People are the foundations of any company. Their full involvement will enable knowledge and experience to be used for the benefit of the company. Education, training, and creation of opportunities for individual growth are aspects that should be taken into account. Typical developmental programs that could be implemented are job rotation, financial incentive schemes, and job training
Process approach	A desired result is achieved more effectively when related resources and activities are managed as a process
Systematic approach	Identifying, understanding, and managing a system of interrelated processes for a given objective will contribute to the effectiveness, efficiency, and efficacy of the approach and company's overall performance
Design improvement and prevention	This principle is primarily applicable at the project level, although the company's internal business processes should have a quality focus. At a project level: the project's management should place strong emphasis on design quality (this is only applicable for nontraditional procurement systems where the contractor has control of the delivery process)—as well as problems and waste prevention achieved through building quality into products, services, and production processes. Typically the costs of correcting problems at the design stage are much lower than later in the process
Factual decision-making	Effective decisions and actions are based on the analysis of information and data. Information and data needed for quality improvement and quality assessments have numerous origins, such as the customer, project performance, competitive comparisons, subcontractors, and suppliers. Through detailed analysis of the information and data acquired, an evaluation to support decision-making at various levels of the organization will be able to take place
Partnership development	The organization seeks to build internal and external partnerships to better accomplish their goals. Internal partnerships include agreements with unions. These agreements will predominantly occur at a project level with site management, for example, restructuring of awards and improvement of site facilities and safety. For external partnerships these will include subcontractors, suppliers, consultants, etc. An increasingly important kind of external relationship is that of a partnering and strategic alliance

2.5 Literature Survey

After the Second World War, most Japanese companies had to start from the ground up. Competition was increasing among different manufacturers. As a result, KAIZEN had become a part of life. It was a fortune that experts like Deming and Juran had introduced the various tools that helped to elevate the KAIZEN concept to new heights in Japan in the late 1950s and early 1960s (Yung, 1996). The relevant literature is classified into three different categories:

- Literature related to conceptual framework
- Literature related to case studies
- Literature related to surveys/empirical research

2.5.1 Literature Related to Conceptual Framework

Kim and Mauborgne (1999) called incremental improvement "imitation" and not "innovation." According to these authors, companies should focus on a proactive strategy, which focuses on the creation of new customers as well as sustaining existing customers. They refer this strategy as "value innovation strategy" where the emphasis is on value and customers and to a lesser extent on the competition. The focus on value innovation pushes managers to go beyond continuous incremental improvements of existing products, service, and processes to new ways of doing things.

Leede and Looise (1999) reviewed some existing organizational designs for CI on these three essential characteristics of CI. As an alternative to the shortcomings of current organizational designs for CI, the authors presented the mini-company concept, related to the socio-technical concept of the self-managing team. The mini-company concept incorporates the three key issues: it has a self-propelling capacity for CI, involving everyone on the shop floor. A constant and market-oriented source for improvement is found in the clients and suppliers of the mini-company. Results of an in-depth case study were presented, showing some strong effects of the mini-company concept.

Bessant, Caffyn, and Gallagher (2001) defined five evolution stages in CI practices in organizations. These stages range from 1 to 5, and the first one (pre-improvement) occurs when the organization introduces the concept of CI without influencing organizational performance. In the fifth level (overall continuous improvement), the whole organization is involved in improving activities of incremental and radical innovations. At this stage, sharing of knowledge and experiences also takes place making it a model of organizational learning.

Williams (2001) highlighted that CI techniques are the recognized way of making significant reduction to production costs. Quality function deployment (QFD) is a well-known technique for translating customer requirements for a product into functional specification. Data suggested that the best opportunity for significant reduction in the overall cost of manufacturing a product is at the design stage of the new product development program.

Bolton and Heap (2002) suggested that CI is an unworkable approach to organizational transformation, that discontinuity is essential to ensure that gains are consolidated and locked into baseline performance. There is a variety of routes to organizational transformation; all must start with an understanding of where the organization wishes to be at all end of the program. However, critically the approach is based on a view that long-term improvement requires periods of consolidation that discontinuity must be explicitly brought into the process.

Chen and Wu (2004) explained that CI can be generated and sustained through the promotion of good improvement model and management support. In fact, it is not so easy in reality. The improvement case may fail without carefully examining the problem in the activity.

Abdolshah and Jahan (2006) described about how to use continuous improvement tools in the different life periods of the organization. Organizations are facing the problem that which CI tool should be used during various stages and different life periods of organization. Methodologies of applying both quantitative and qualitative tools in the different life periods of an organization were discussed.

Chakrborty, Bhattacharya, Ghosh, and Sarkar (2013) defined KAIZEN as a constant effort not only to maintain but also to upgrade standards. KAIZEN pronounces perpetual development in all walks of our life. This is an inborn instinct present in every human being. Japanese are very diligent by nature though they sometimes are not even cognizant of KAIZEN concept. The rate of KAIZEN development is very low but it gradually brings about a great development during the course of time, which is very much helpful for the companies to sustain.

Dhongade, Singh, and Stouty (2013) argued that KAIZEN tools such as Poka-Yoke, 5 "S" Concept, and 7 kind of Productive Loss are used to increase the productivity of the process in the form of continuous improvement. The effective implementation of KAIZEN methodology will lead to the success of the organization. This paper discussed the literature that has been published in this field and presents a review of literature which will be helpful to new research in this field. The authors also feel that Poka-Yoke is the most important tool which will help in full proofing of any deviation in the system.

Jonsdottir, Ingason, & Jonasson (2014) have analyzed how certified organizations support continuous improvement: whether continuous improvement is handled like a formal project; who is responsible for carrying out continuous improvement projects; and which project management tools are mainly used for these projects. It is concluded that action supporting CI of certified organizations are not handled properly as a part of CI project. Part of CI projects could be "just-do-it" actions and no need to use formal PM (project management).

2.5.2 Literature Related to Case Studies

The different case studies performed by different researchers have been discussed below:

Watson (1998) discussed the application of CI techniques used by Tracor Aerospace Company. The author considered elements such as leadership, strategic planning, and customer and market focus. Data and information were collected continuously, measuring and monitoring all aspects of the business. In conclusion, Tracor team members are very proud of their accomplishments and are dedicated to continuing and improving all our methods. We truly aspire to be world class in managing our business, to be viewed as the number one supplier for selected defense products, systems, and commercial derivatives by offering proven, low-risk technical solutions combined with the best understanding of our customers' problems, just as our Vision statement says. Manufacturing processes can be managed using data and SPC (statistical process control) and are continuously improved to ensure that results meet all goals and expectations.

Bessant and Francis (1999) described that for developing CI capability, organizations need to move to a level of development in which strategic goals are communicated and deployed and where improvement activity is guided by a process of monitoring and measurement against these strategic objectives. Policy deployment of this kind is more prevalent in Japanese examples and in a handful of cases in Western firms. Implementing it poses significant challenges and requires a different and additional toolkit of enabling resources. This study reported on the experience of policy deployment in Japan and in Western enterprises and explores some of the implementation issues raised. The paper had briefly outlined some of the characteristics of organizations deploying CI for strategic advantage.

Adams, Componation, Czarnecki, and Schroer (1999) explained that simulation is the powerful tool to support CI process improvement. Two case studies including a commercial manufacturer and aerospace manufacturer were performed where simulation is used to support the continuous improvement steps. In summary, the following conclusions are made:

- Process simulation can be used to support steps in the CI process.
- To be most effective, simulation model should be developed.
- For new situations, basic, simple model of the process is a good way to start.
- Interpreting the results with management can be beneficial.

Savolainen (1999) conducted two case studies including a medium-sized metal industry and other larger group in the construction and concrete industry. The main aim of the studies was to increase the understanding of the processes and dynamics of continuous improvement implementation. Focus is placed on how these companies are renewed through the embedding of quality-related management ideology. The paper discussed the processes and dynamics of CI implementation conceptually and empirically. Results show that the dynamics of CI implementation process is cyclic in nature, which progresses at different speeds and with varying intensity.

Burns (2000) described the importance of two techniques, namely overall equipment effectiveness (OEE) and setup reduction, taking an example of Weston EU Company. No appropriate measures of the process and equipment usage are available. Initially, six pilot areas have been identified; out of these three turned out to be successful. OEE is actually used to drive CI in the development of a company. Setup reduction was applied to reduce change over times, to meet the customer demand for greater product mix, and to overcome the difficulties in machine loading. Both techniques are described in terms of how they help the company to drive improvement in the core of the business—70 capital equipment CNC machines.

Lee (2000) conducted a case study at Nichols Foods manufacturing food products. There was a lack of standard operating procedures, forces, and structure. The study describes how the company values have improved the work environment for the employees and motivated them to achieve excellence. The paper described how the KAIZEN program was implemented in this company using 5S technique and team training. The result showed the decrease in quality rejections, reduction in change over times, and increase in manufacturing efficiencies.

Lee et al. (2000) described that inclusion of the KAIZEN approach in industrial technology is beneficial. This case study provided a description of the steps used to implement lean thinking in a typical Midwestern company developing a dynamic tri-resin fiberglass rod, which has hundred times more tensile strength than that of steel. After implementing lean thinking, reduction in space used in the building and material handling costs and also lower scrap rates can be expected. Exercises that are described in this paper can be used within existing system in all manufacturing-focused programs to assure that graduates are sufficiently familiar with this important thrust.

Palmer (2001) focused his study on the "Inventory management KAIZEN" that has been conducted at "BAE SYSTEM" to remove the muda (waste) from the receiving and storing process. KAIZEN event encompasses about 5 months, 1 week from actual analysis of the process and the remainder working to implement the changes that are identified. Results indicated that the process time was reduced from 610 to 290 h. Ultimately the KAIZEN event results in saving well over million dollars per year.

Ahmed, Hassan, and Fen (2005) conducted a study in a casting-based manufacturing plant, which is currently implementing KAIZEN towards achieving higher productivity. The study has focused on the performance indicators (PIs) currently being measured by the company. Careful investigations and observations were taken to show the effectiveness and efficiency of the implementation of the KAIZEN system in an innovative manner. After analyzing the collected data, sufficient information was generated on various aspects of performance evaluation. However, due to lack of financial data, monetary-based PIs could not be carried out in this study.

Granja, Picchi, and Robert (2005) studied the target and KAIZEN costing concept in the construction-based company. The aim was to develop the framework taking together these two matching approaches, which provides a basis for a total cost management system. The authors explained that the continuing series of KAIZEN activities are needed to achieve product performance and reduce the cost. Combining target and KAIZEN costing is a powerful approach for construction company by assuring value for the customer at a low but profitable price.

Kikuchi, Kikuchi, and Takai (2007) aimed at applying overall equipment effectiveness (OEE) method to cost reduction by using KAIZEN technique in a semiconductor industry. The consumption of gases and chemicals (GAC) for a specific process was very high. Two different methods of overall consumable effectiveness (OCE) technique were adopted to reduce the consumption of gases and chemicals (GAC) for 12 items. Result indicated a cost reduction of 7% annually for GAC usage. This experience had raised the engineer's awareness that the KAIZEN process can be applied to other areas also.

Chandrasekaran, Kannan, and Pandiaraj (2008) applied KAIZEN technique to solve the "part mismatch problem" in automobile assembly production line. Step by step KAIZEN procedure was followed to solve the problem by data collection, root cause analysis, selection of the best solution method, corrective action, and documentation. The various benefits that have been observed after implementing KAIZEN include elimination of major functional problem, reduction in quality rejections, elimination of rework processes, and a considerable cost savings.

Jung, Wang, and Wu (2009) explored the relationship between competitive strategy, total quality management (TQM), and CI of international project management (CIIPM). Based on a literature review, a theoretical model and five hypotheses are developed. A cross-sectional data set collected from 268 international project managers based in four countries is used to test the theoretical model. The results from the statistical analysis suggested that competitive strategy does not directly influence the CIIPM performance, but it influences through the mediation of TQM practices. The results also suggested that "human resource-based" TQM elements have stronger influence towards CIIPM performance than "technology-based" TQM elements. TQM elements fully mediate the relationship between competitive strategy and CIIPM. This implies that organizations need an innovative management methodology, such as the TQM practice, in order to achieve competitive strategy materializing towards international project management performance. Further, the findings suggested that it is the soft TQM elements (i.e., top management's leadership, compensation, training, and empowerment) that impact the CIIPM more significantly.

Yoo et al. (2012) investigated both internal and external effects of defective production and delivery from imperfect production and inspection processes in a stable production and inventory system, and subsequent defective returns and dispositions. The authors presented the models under continuous improvement investments over multiple periods, which have not been explored before. They developed an algorithm for inspection option change to determine the inspection alternative yielding a superior cumulative profit over multiple planning periods, among inspection option change, entire lot screening, and no inspection.

Ionica, Leba, and Corbu (2013) described that reengineering and continuous improvements are considered extreme approaches in the "life" of an organization. The present study underlined both the major differences and the resemblances between them. Researchers bring the meaning of these two approaches into the personal life development. The orientation towards one of them depends on several individual characteristics. The authors applied several tests that deal with personality, the "taste" for risk, self-fulfillment through career, adaptability to changes, etc. These tests were implemented in a software system. The results pointed the correspondences between the life stage and the suited personal development model.

Kaskhedikar and Katore (2013) suggested that KAIZEN management tools are the key elements for any manufacturing firm and are essential to study for providing a better quality management in industries. This can be achieved by developing an instrument which will consist of the critical success factors and performance measurement factors, which are responsible for overall growth of casting industries.

Savino and Mazza (2014) have conducted a study that provides some general principles to integrate environmental management system (EMS) in quality management system (QMS) development and suggested an implementation of continuous improvement, which addresses, analyzes, prioritizes, and measures activities within the Deming wheel. Under industrial management perspective this study provides an interesting opportunity to evaluate and to implement the Deming wheel, suggesting a practical outline of activities that should be conducted within the PDCA cycle. This study suggested that the following future practical investigations are needed: an assessment of the core resources for continuous improvement, classification of continuous improvement capabilities of firms, and how to measure the attitudes of firms for continuous improvement.

2.5.3 *Literature Survey Related to Surveys/Empirical Research*

The survey performed by different researchers has been discussed below:

Gibb and Davies (1990) identified and highlighted the critical success factor for CI and innovative strategy in Australian SMEs, importance of market orientation, and effective strategic formulation in successful SMEs. The critical success factors that were highlighted in the survey include promoting a corporate culture, creating an effective structure, analyzing competitors, developing cooperations and partnerships, and developing flexibility and speed of response.

Chang (1995) discussed a recent survey of companies which shows that although 73% of the respondents had implemented total quality initiatives, 63% of these initiatives were failing. It was suggested that many organizations are trying to implement too many quality improvement activities too fast without taking the time to develop systematic, long-term implementation of strategies. The author also provided the guidelines on critical success factors for CI and their impact on management, empowerment, learning, and training.

Soderquist (1996) investigated CI and innovation practices in French SMEs. In this survey, they examine the drivers for change and the short- and long-term goals, the sources of innovation, and the nature of innovation management in French SMEs. Respondents were asked to consider a recent and successful innovation in product and then to indicate just how important a number of items are used as a source of particular innovation. The top nine sources of innovation that have been found include introduction of the new product, continuous improvement of work processes, radical change (e.g., through business process reengineering), increased focus in marketing/sales efforts, reduction in indirect staff numbers, improvement on staff competence, improved quality of product and services, improving the quality of management, and efforts to improve supplier performance. The survey identified two groups of SMEs. The first group has reported satisfaction with their organization's performance in product innovation and has also reported that their organizations have a strategic approach to innovation. The second group comprises SMEs, which are satisfied with current actions for improving short-term performances. Further analysis showed that the second group is more likely to report a stronger emphasis on performance management approach.

Chapman and Hyland (1997) examined the results of work being done by a team of Australian researchers on CI in Australia. The research is part of a coordinated international effort involving the UK-based Continuous Improvement Research for Competitive Advantage (CIRCA) project and the European Continuous Improvement Network (EuroCINet). The paper reported on CI activities in 385 responding units located across the major manufacturing states in Australia. Results indicated that CI is a long-term process that facilitates ongoing change in manufacturing organizations throughout the world. Those companies which had the courage and foresight to implement CI strategies will be better prepared to deal with threats from competitors both domestic and overseas. CI had proven to be a highly effective process for achieving these and other goals.

Irane and Sharp (1997) suggested, based on the survey in a small-scale manufacture, that continuous improvement (CI) strategy should be engrained as a belief into the employees' heart. The ideal situation of CI strategy is its integration with the corporate culture.

Bessant (2000a) presented a survey that has been conducted by CI research advantage (CIRCA) at UK firms. Survey suggested that 65% of companies consider continuous improvement (CI) to be of strategic importance, around 50% have instituted some form of systematic program to apply these concepts, 19% claim to have a widespread and sustained process of CI in operation, and of those firms using CI 89% claim that they had impact on productivity, quality, delivery performance, or combination of these.

Hongming, Sun, and Xu (2000) carried out a survey in Chinese companies and found that not all companies that have carried out CI activities achieved desired results. It has significant impact on companies in which CI implementation requires adequate input on company capital human resource and organizational activities. It is a challenge for companies in the organizational structure business principle and operation methods. Hyland, Mellor, Omara, and Kondepndi (2000) examined five small to medium manufacturers and used a mapping tool that measures the extent of learning within the firms. If firms using CI are to fully benefit from the learning process then they must have a strategy in place that ensures that knowledge is captured and the workforce is willing to transfer knowledge throughout the organization. It was concluded that the effectiveness of a CI program in an organization is related to the level of integration between the CI process and the learning within the organization. CI requires a long-term commitment to a course of action and the development of a set of beliefs.

Mackle (2000) presented a survey conducted by KAIZEN institute that has been designing and implementing various continuous improvement programs in most of the companies in the UK. Institution has conducted a survey with all of their UK clients. Outcomes of the survey showed that organizations have not successfully implemented these improvement programs. The opportunities for improvement were also identified in this survey.

Terziovski (2001) presented the result of a mail survey used to investigate the relationship between CI and innovation practices and SME performance in 115 Australian manufacturing industries. A total of 19 questions were included in the questionnaire. 57 independent variables and 12 dependent variables were analyzed using multi-regression analysis. The author concluded that a continuous improvement and innovation management strategy and system are significant predictors of SME performance.

Gonsalves (2002) performed a survey about the effect of ERP and CI on the performance in 500 manufacturing companies. The author concluded that CI implementation has positive influence on business process reengineering (BPR) execution. Integrated CI and BPR have positive effects on the company's performance.

Teng, Ho, Shumal, and Liu (2006) investigated the effects of CI and cleaner production on the operational performance. A total of 223 responses were obtained after the distribution of questionnaire. Sample for study has been collected via a survey of Taiwan electronic manufacturing firms. The direct and indirect influences of independent variables on dependent variables are tested by structural equation modeling (SEM) technique. The results showed that the continuous improvement might not be able to directly improve the operational performance. However, CI plays a significant role in cleaner production implementation.

Yan-jiang, Lang, and Xiao-na (2006) conducted a survey by using data of the global continuous innovation network to analyze the influencing factors of CI. This survey designs 18 questions to describe the reasons why companies are implementing continuous improvement activities, 13 questions to describe the company's external environment, and 11 questions to describe the situation of continuous improvement activities in functional departments of the companies. Results showed that the internal motivation factors are responsible for popularization of CI activities and have varying degrees of influence on these activities.

Terziovski and Power (2007) tested the strength of CI in finding out the relationship between motivation for seeking ISO 9000 certification, quality culture, management responsibility, and perceived benefits derived from ISO 9000 certification. The research involved the development and application of a survey instrument applied to a sample of 1500 ISO 9000-certified organizations. The response rate was 27% (400 responses). The data was analyzed using SPSS for Windows, and multivariate functions. Among the major findings is that organizations that seek ISO 9000 certification with a proactive approach driven by a continuous improvement strategy are more likely to derive significant business benefits as a result. The authors also found that organizations can effectively use ISO certification as a means of promoting and facilitating a quality culture, where the quality auditor is an important player in the continuous improvement.

Sim and Rogers (2009) focused their survey on a 500 manufacturing plant located in the Eastern USA. A survey was distributed to both salaried and unionized hourly employees. The purpose of this survey was to understand why implementing

CI strategies can be difficult at times. It also addresses the problem of resistance to change within even those firms whose CEO is most fully committed to implementation of CI programs. Results from the survey indicated that the problem lies primarily with an aging and high-seniority hourly workforce and a lack of committed leadership at this research site. For example, salaried employees consistently provided higher positive ratings of CI initiatives. In addition, higher seniority was directly correlated with negative ratings. Finally, the study found that employees do not feel valued when they contribute to the improvement processes and that 100% of the hourly male employees disagreed, "The Company considers the employees as the most important asset and will do what ever they can to keep their people."

Jaca, Viles, Mateo, and Santos (2012) aimed at evaluating the importance of the factors reported in the literature as enablers of CI programs and determining the perception of managers of different companies in the Basque Country and Navarre (Spain) regarding the relevance of these factors to their improvement programs. 15 elements were considered to be key issues for the sustainability of CI programs. Semi-structured interviews were conducted with 36 companies in order to assess how the companies value the factors and how the factors are applied and measured. The findings regarding the application and evaluation of such factors had revealed that companies are focused on the agents associated with the achievement of results. Other factors, such as management commitment or the promotion of team working, are highly scored and applied, but few companies evaluate them or take actions to improve their application. The paper had analyzed the application of some factors considered to be enablers or key factors for the sustainability of continuous improvement systems. Furthermore, it examined the mechanisms or indicators which are used by some companies to measure the application of those factors.

Oprime, Mendes, and Pimenta (2012) identified and analyzed the critical factors in the development of CI activities in 46 Brazilian companies. Conceptual models of the relationship between practices and results have been tested to identify the critical factors using a survey conducted in 46 industrial companies. Nonparametric tests have been used to test the hypotheses developed based on the literature. These critical factors are related to actions that encourage employees to participate in CI activities and incentive mechanisms to be able to apply identification techniques and tools successfully, as well as find solution to problems. The results indicated the importance of staff training in problem solution tools, incentives for suggestions, face-to-face communication, and regular shop floor visits such as critical factors for the success of continuous improvement activities. Operational practices of CI contribute to company performance in relation to improvements in productivity, quality, lead time, cost, customer satisfaction, and development of employees' skills to solve problems.

Holtskog (2013) investigated the historical line of continuous improvement in one of Norway's most important automotive industry clusters, where continuous improvement has been an issue over decades. A survey of 600 respondents showed that continuous improvement is felt like a natural part of the tasks in daily work-life. The companies in this survey had done this in many ways and over time shifted their way. Basically, there are four categories: education in how to do thing, simplifications of operations or developing a "common sense" approach, close involvement with the unions, and close collaboration (both formal and informal) of leaders and employees. Education is about learning the tools, but it also makes each employee capable to work independently and take autonomous decisions. Cultural aspects of the work and how people look at themselves are important when local adaptation shall be done. And finally unions can play an important role in the improvement work, thereby helping the company to stay profitable.

2.6 Benefits of Implementing CI Approach

Different authors have discussed the benefits of implementing CI approach as discussed below:

KAIZEN approach aimed at 89% improvement in work in process (WIP) and 88.5% increase in productivity and floor space requirements were saved over 2000 sq. ft. in Allied Signal Inc. jet engine manufacturing industry (Sheridan, 1997); increase in the production rate of about 80% and drop in the error rate from above 50% to about 1% at fuel-fitter assembly (Erlandson, Noblett, & Phelps, 1998); decrease in 11% stations, decrease in 11.7% moving around, 16% time saving, 34.2% decrease in length, and saving of 53% transportation cost at National Productivity Improvement Program (NPIP), at Chaharmahal-Bakhtiari Agriculture Organization (Dehghan, Zohrab, Momeni, & Hoseini, 2006); improving safety aspects and better communication at UK-based multinational chemical company (Vassie, 1998); improving the performance of manufacturing operations and practical evaluation of continuous improvement tools in a leading company of Korea (Leem & Kim, 2004); organizational development, cultural change, and higher involvement of employees in improvement activities in a mining company of South Africa (Jager, Minnie, Jager, Welgemoed, Bessant, & Francis, 2004); reducing failure cost by 10% and reducing waste due to waiting at a company in Brazil (Santos, Powell, & Formoso, 2000); helped in environmental management, total employee involvement, and improved purchasing at Housing Action Trust in Waltham (Tomlinson, 1994); improving organization structure, effectiveness of equipment, and momentum of the incremental improvements (Cook & Dale, 1995); planning and execution of small incremental changes (Jha, Michela, & Noori, 1999); improved employee involvement, improving the culture of continuous improvement, and improving the companies' competitiveness (Garcia-Lorenzo, Prado, & Arca, 2000); improving the working environment and efficiency of the employees (Tonnessen, 2005); improving learning ability, developing core competencies, and encouraging team-based improvement (Ljungstrom, 2005); Woods (1997) stated that CI benefited employees (by providing a healthy workplace), satisfied customers, and increased financial returns for the company; reducing lead time by 23% and setup time by 39% at Australian Wiring Systems (Rusiniak, 1996); Pentair reported a savings of 112,000 sq. ft. through KAIZEN approach (Siekman, 2002); work area transformation and employee development (Farris, VanAken,

Doolen, & June, 2009). Williams (2001) highlighted the major potential benefits of CI approach. These benefits include:

- Increased business performance (in terms of reduced waste, setup time, breakdowns, and lead time)
- Increased "people performance" in the form of improved development
- Empowerment
- Participation
- Quality of work-life of employees, all of which address contemporary societal needs

According to (Khan, 2011), KAIZEN reduces waste in areas such as inventory, waiting times, transportation, worker motion, employee skills, overproduction, excess quality, and processes. It improves space utilization, product quality, use of capital, communications, production capacity, and employee retention. It also provides immediate results. Instead of focusing on large capital intensive improvements, KAIZEN focuses on creative investments that continually solve large numbers of small problems. Large capital projects and major changes will still be needed, and KAIZEN will also improve the capital project process, but the real power of KAIZEN is in the ongoing process of continually making small improvements that improve processes and reduce waste.

2.7 Concluding Remarks

From the literature survey described here, it is seen that the general consensus of CI approach and strategies is a very effective manufacturing philosophy across the globe. These strategies are universal in nature and encompass all aspects of manufacturing. These include the generation of an appropriate cultural environment, which includes total commitment at all levels of management and cooperation within the workforce. It is concluded that there is reasonably vast literature available on CI approach and strategies, which gives a broad view of past practices and researches carried over the globe.

Chapter 3
Design of Study

3.1 Introduction

This chapter deals with the design of the study, which includes different phases of research work. The details of the work done in each phase along with the tools, techniques, and models used have been explained.

3.2 Problem Formulation

Today, the ultimate objective of the SMEs is to increase productivity through simplification of manufacturing system processes and incremental improvements by using modern techniques like continuous improvement (CI), total productive maintenance (TPM), and total quality management (TQM). Small–medium-sized enterprises (SMEs) are the important parts of modern economies. Over the years competition among these enterprises has been increased, which results in the survival of the SMEs to a large extent. To compete in this continuously changing environment, these companies must identify new methods allowing them to remain competitive, enabling their companies to respond rapidly to new demands. Competition and continuously increasing standards of customer satisfaction have proven to be drivers of organizational performance improvement. In order for SMEs to remain competitive, retain their market share, meet these continuously changing needs, and satisfy customers, continuous improvement of manufacturing system processes has become necessary. Keeping in view the above-mentioned facts, different strategies will be involved in successful implementation of continuous improvement approach in various SMEs. This study is an attempt to enhance the performance of SMEs using continuous improvement strategies and the impact of these strategies on SMEs of Northern India.

J. Singh, H. Singh, *Strategic Implementation of Continuous Improvement Approach*, SpringerBriefs in Operations Management,
https://doi.org/10.1007/978-3-319-93121-0_3

3.3 Phases of Research

The research design comprises five phases: literature review, survey of the manufacturing industry by employing strategically designed questionnaire, conduction of case studies in different organizations, inferences drawn from the survey and case studies that have been synthesized for the development of implementation plan, and structural equation modeling applied to various initiatives of CI approach.

- Phase one: Literature review provides deep insight into different CI strategies and approach. Research gaps have been found earlier in Chap. 2.
- Phase two: The survey of various mechanical enterprises has been conducted through a specially designed questionnaire for understanding and assessing the prevailing situation in the Indian industry. In this questionnaire, objective-type questions containing open ended and closed ended have been framed. The survey has been conducted mainly through mail but personal visits have also been made to have good responses. Statistical analysis of the primary data collected through the survey has been carried out to assess the following issues:

    1. Level of usage of continuous improvement strategies utilized by manufacturing organizations: What is the present status of use of different CI strategies in Indian SMEs?
    2. To assess the importance level of these strategies.
    3. Important barriers and key enablers and performance of these key enablers have been identified.
    4. To assess the performance of continuous approach in the form of benefits.
    5. Different issues related to continuous improvement approach have been analyzed.
    6. Ranking of the strategies has been done using AHP and VIKOR through specially designed questionnaire.

    The various statistical analysis tools like Cronbach's alpha, Pearson correlation coefficient, two-tailed "*t*" test, *z*-test, and multiple regression analysis, paired sample "*t*" test have been employed in the research to evaluate the contributions of different strategies and enablers towards improving manufacturing performance.

- Phase three: The survey has been followed by case studies in some industrial units. The purpose of the case studies is to analyze these strategies of CI approach and check the performance of these strategies.

3.3.1 *Identification of Industrial Units*

Case studies have been conducted at the following organizations: M/S Highway Industries Limited, Ludhiana, and M/S Farm Parts Limited. The industrial units selected for conduction of case studies have been selected on the following bases:

- The units to be representative of the manufacturing industry, i.e., most of the operations and processes are prevalent in that unit.

- It has requisite activity going on pertaining to the potential areas as reflected by the results of the survey.
- The unit is forthcoming and cooperative for conducting case studies.
- Case studies in one small-scale and one medium-scale industries have been conducted.

- Phase four: The information obtained from the case studies as well as from survey part has been synthesized.

3.3.2 Options Field Methodology

- To list the various options as solutions.
- To make categories from options.
- To decide the dimensions of the problem.
- To cluster and order the dimensions.

3.3.3 Options Profile Methodology

- To make the various profiles as a solution to the problem.
- To draw a tie line.
- To allocate various options to the profiles.

3.3.4 Analytic Hierarchy Process

- To enlist the various criteria as objectives.
- To do paired comparison of objectives and assigning numerical values.
- To calculate the weighted sum of objectives by Eigen vector.
- To find comparative contribution of each profile towards each objective.

The analytical hierarchy process (AHP) is a decision-aiding method developed by Satty. It aims at quantifying relative priorities for a given set of alternatives on a ratio scale, based on the judgment of the decision-maker, and stresses the importance of the intuitive judgments of a decision-maker as well as the consistency of the comparison of alternatives in the decision-making process (Al-Harbi, 2001). In using the AHP, one constructs a hierarchy (consisting of goal, criteria, and alternatives), and then makes judgments (or performs measurements) on pairs of elements with respect to a controlling element. Ratio scales are derived from these judgments and then synthesized throughout the structure to select the best alternative. Decision-making involves criteria and alternatives to choose from. The criteria usually have different importances and the alternatives in turn differ in our preference for them on each criterion. To make such trade-offs and choices we need a way to measure. Measuring needs a good understanding of methods of measurement and different scales of measurement (Bahurmoz, 2006).

3.3.5 Structural Equation Modeling

- Preparation of model describing the latent and measurement variables.
- Combination of exogenous and endogenous variables.
- Determining the path coefficient for the prepared model.

- Phase five: Finally, a structured development plan has been suggested to act as the guideline for the manufacturing industry for improving the SME performance using CI strategies.

3.4 Overall Equipment Effectiveness (OEE)

The overall equipment effectiveness has been originated from the total productive maintenance practices, developed by Nakajima (Sharma, Kumar, & Kumar, 2006).

OEE is equal to the multiplication of the three main parameters as shown below:

- Availability indicates the problem which is caused by downtime losses.
- Performance indicates the losses caused by speed losses.
- Quality indicates the scrap and rework losses.

TPM can be defined in terms of overall equipment effectiveness (OEE) which, in turn, is a function of availability (A), performance efficiency (PE), and quality rate (QR):

$$\text{Availability} = \frac{\text{Loading time} - \text{Down time}}{\text{Loading time}}$$

$$\text{Performance efficiency (PE)} = \text{Operating speed rate} * \text{Net operating rate}$$

$$\text{Operating speed rate} = \frac{\text{Theoretical cycle time}}{\text{Actual cycle time}}$$

$$\text{Net operating rate} = \frac{(\text{Processed amount}) * (\text{Actual cycle time})}{\text{Operating time}}$$

$$\text{Quality rate}(\text{Q}) = \frac{\text{Processed amount} - \text{Defect amount}}{\text{Processed amount}}$$

OEE = (Availability)* (Performance efficiency)*(Quality rate) (Blanchard, 1997).

3.5 Objectives and Issues of the Study

3.5.1 Objectives of the Study

- To assess the present status of CI strategies in SMEs.
- To implement CI strategies in selected companies.
- Developing a model for improving the performance of the manufacturing industry.

3.5.2 Issues of the Study

- To assess the performance of initiatives taken to improve the performance of SMEs.
- To ascertain the performance of CI strategies and approach for strategic success.
- To assess the performance of different enablers of CI approach.
- Development of a conceptual model for performance improvement.
- Preparing the implementation plan of the developed model for improving the performance of SMEs using continuous improvement strategies.

3.6 Overall Methodology

The research is aimed at developing an understanding of CI approach. The methodology comprises five phases, viz.

- Exhaustive literature review.
- Survey of the mechanical industry by employing strategically designed "questionnaires."
- Conduction of case studies in different organizations.
- Inferences drawn from the survey and case studies have been synthesized.
- Development of implementation plan.

The literature about various phases of CI implementation has been reviewed to clarify the context as well as to frame questions to be included in a questionnaire for conducting a survey of the mechanical industry. The literature review has been done to eliminate waste for achieving excellence. The survey has been carried out to get the information on various aspects of strategic implementation of CI approach. Based on the survey, a relationship has been established between various input and output parameters. The various statistical analysis tools like Cronbach's alpha, Pearson correlation coefficient, *z*-test, multiple regression analysis, two-tailed "t" test, and paired sample "t" test have been employed to evaluate the contributions of CI strategic implementation towards building manufacturing performance improvements through different benefits. The tools used for multi-criteria decision-making (MCDM) including AHP and VIKOR have been employed so as to rank the different strategies used in the research. Case studies have been undertaken to closely study the various aspects of implementing CI strategies. The findings of the survey and results of the case studies have been synthesized and a management plan has been presented to improve the performance of SMEs for strategic success. Furthermore, structural equation modeling helps to predict the dependency of one CI initiative over other and determine the standard path coefficient for prepared model.

To authenticate the data obtained through questionnaire, case studies have been conducted in the key manufacturing units, viz.; M/S Highway Industries Ltd.; and M/S Farm Parts Limited.

Chapter 4
Analysis of Preliminary Data

4.1 Introduction

This chapter presents a detailed survey conducted in various organizations in the northern part of India. The objective of the survey is to determine the present status of continuous improvement strategies in the Indian manufacturing industry, to assess the important barriers that affect the implementation of CI strategies, and to assess the role of key enablers leading to improve the performance of manufacturing operations for strategic success.

4.2 Methodology

The methodology employed in the study has been depicted in Fig. 4.1.

4.2.1 Creation of Industry Database

A database of 500 small- and medium-scale organizations of Northern India SMEs has been created, which contains name of company; location; main products; type of industry; and their postal addresses. The target companies have been selected from the following two sources:

- The Confederation of Indian Industry (CII)
- Directorate of Industries

J. Singh, H. Singh, *Strategic Implementation of Continuous Improvement Approach*, SpringerBriefs in Operations Management,
https://doi.org/10.1007/978-3-319-93121-0_4

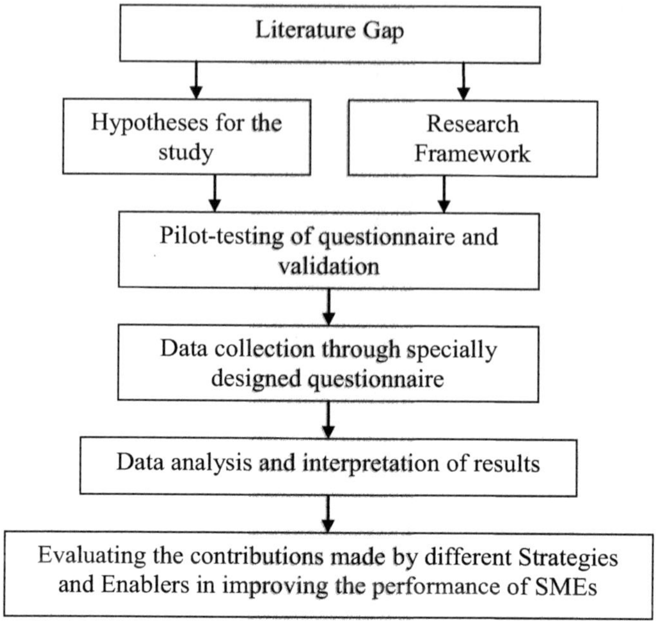

Fig. 4.1 Methodology used for the research study

4.2.2 Pilot Testing of Questionnaire

To ensure the relevance and the trueness of the questions to the mechanical industry, the questionnaire has been pilot tested on a representative sample of industry. The suggestions from the peers, consultants, senior executives from the industries, and academicians have been incorporated to make the questionnaire relevant to the purpose and bring out key outcomes. In order to ascertain the contributions of CI approach towards realization of performance improvement, a detailed "questionnaire" has been designed for assessing the manufacturing performance accrued by Indian SMEs as a result of strategic CI implementation. The questionnaire survey technique has been deployed in the present study for seeking information on status of CI strategies and realization of various manufacturing performance in the Indian mechanical industry. For effectively conducting the survey, the CI questionnaire has been designed through extensive literature review and validated through peer review from academicians, consultants, CI councilors, and practitioners (CI co-coordinators) from the industry. Figure 4.2 depicts the steps undertaken in finalization of the specially designed questionnaire.

4.2.3 Filling of Questionnaire

A blank questionnaire designed on 4-point Likert scale is used for the survey. A final structured survey questionnaire has been mailed to 300 members chosen at random from among the membership of the Confederation of Indian Industry (CII)

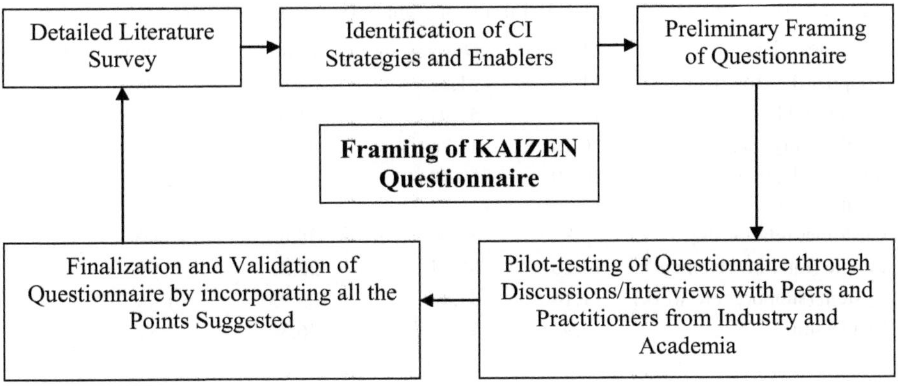

Fig. 4.2 Framing of KAIZEN questionnaire

Table 4.1 Breakdown of responses by the organizations

Type of company	Small scale	Medium scale
	53 53/101 = 52.48%	48 48/101 = 47.52%

and Directorate of Industries. Subsequently, reminders were sent to all the firms to expedite the matter. Information from some of the organizations was received in this way. Most of the information, however, has been collected by making personal visits to the organizations and by having discussions with the executives at the senior levels. The CI questionnaire serves the purpose of revealing exploits of Indian entrepreneurs with CI practices and highlights the contributions of CI in realizing the overall organization's goals and objectives. In this study, a reasonably large number of manufacturing organizations (101 organizations representing response rate of 33.67%) have been extensively surveyed, to ascertain contributions made by CI approach in the Indian industries towards realization of business excellence. The breakdown of responses is shown in Table 4.1.

Most of the respondents to "Questionnaire" are top brass of management executives including several managing directors, partners, proprietors, managers of different departments, senior engineers, engineers, management representatives, heads of different departments, secretaries, etc. The responses thus received have been compiled and analyzed critically to ascertain the performance of the Northern Indian Industry regarding various CI-related issues.

4.2.4 Hypotheses for the Study: From the Literature the Following Hypotheses Have Been Framed

- H1: There exists overall association between CI strategies and manufacturing performance parameters.

Table 4.2 shows the percentage level of usage of eight (08) main CI strategies

CI strategy/level of usage	Mature (%)	Developing (%)	Immature (%)	Not applicable (%)
Supplier development	11.55	51.82	34.65	1.98
Process (JIT)	13.53	52.97	31.02	2.48
Total quality management	20.79	43.76	32.28	3.17
Total productive maintenance	20.59	50.10	26.34	2.97
Customer relationship	13.86	44.06	38.62	3.46
People (total involvement)	14.26	56.44	28.31	0.99
System (support core work)	11.14	57.43	29.95	1.48
Leadership	7.92	53.47	36.30	2.31

- H2: CI enablers and approach significantly aimed at improving manufacturing performance.
- H3: Improvement in manufacturing performance is a function of experience gained by manufacturing organizations over an extended time period.

4.3 Level of Usage of CI Strategies

The questionnaire filled by manufacturing organizations applying continuous improvement strategies has been analyzed at (08) main strategy level as well as sub-strategies level in terms of (08) main CI strategies or sub-strategies. The percentage usage level has been calculated for all strategies and sub-strategies of continuous improvement approach. Level of usage of 08 main CI strategies is shown in Table 4.2.

4.3.1 Discussion of Level of Usage of Eight (08) Main CI Strategies

The manufacturing organizations of Northern India utilize total quality management strategy to its high maturity level followed by total productive maintenance, people (total involvement), customer relationship, process (JIT), supplier development, system (support core work), and leadership. These organizations have implemented system (support core work) strategies to its highly developing stage than other strategies of continuous improvement approach. Strategies of customer relationship are highly immature in their use and have high level of non-applicability in their use as compared to other organizational strategies of continuous improvement approach (Table 4.3).

Table 4.3 Percentage level of usage of strategies in terms of (08) main CI strategies

Sub-strategy/level of usage	Mature (%)	Developing (%)	Immature (%)	Not applicable (%)
Supplier development				
Total cost management	17.82	69.31	11.88	0.99
Value analysis	11.88	40.59	45.55	1.98
Value stream analysis	4.95	45.55	46.53	2.97
Process (just in time)				
Cycle time reduction	35.64	52.48	10.89	0.99
Process flow analysis	12.87	57.43	25.74	3.96
Material flow analysis	11.88	49.51	37.62	0.99
Theory of constraints	6.93	50.5	39.6	2.97
Redesign	5.94	54.46	37.62	1.98
Workflow analysis	7.92	53.47	34.65	3.96
Quality (TQM)				
5S	65.35	27.72	5.94	0.99
Error proofing analysis	14.85	69.1	11.88	3.96
Failure mode effect analysis (FMEA)	14.85	34.65	49.51	0.99
Six sigma	4.95	50.50	39.60	4.95
Standardization	3.96	36.63	54.46	4.95
Total productive maintenance				
Preventive maintenance	51.49	39.60	6.93	1.98
Autonomous maintenance	28.71	55.45	12.87	2.97
Equipment restoration	10.89	47.53	37.62	3.96
Changeover reduction: SMED	4.95	51.49	38.61	4.95
Minor stoppage elimination	6.93	56.44	35.64	0.99
Customer relationship				
Quality function deployment	14.85	64.36	17.82	2.97
Customer quality, cost, delivery analysis (QCD)	12.87	23.76	59.41	3.96
People (total involvement)				
Internal training and monitoring	8.91	84.16	6.93	0
Manager development	19.80	4.65	41.59	3.96
Suggestion system	21.78	53.47	23.76	0.99
Team-based improvement	11.88	50.50	37.62	0
Principles of KAIZEN	8.91	59.41	31.68	0
System (support core work)				
Finance	14.85	68.32	16.83	0
Marketing	14.85	44.56	38.61	1.98
Process flow mapping	6.93	63.37	25.74	3.96
Support and administrative KAIZEN	7.92	53.47	38.61	0
Leadership				
Vision alignment and direction	7.92	67.33	24.75	0
Policy deployment	10.89	36.64	47.52	4.95
Recognition	4.95	56.44	36.63	1.98

4.4 Level of Usage of Sub-strategies of CI Approach

4.4.1 Discussion of Level of Usage of Sub-strategies of CI Approach

The results of investigation demonstrated that managing the financial outcomes of all the activities is highly mature in use to work continuously with suppliers; reducing the cycle time of the operations is highly mature in their use to deliver the product with right quality, right time, and right quantity to the customers; organizing a workplace by 5S activities is highly mature in their use in terms of total quality management; preventing the failure before it occurs is highly mature in their use in terms of total productive maintenance; translating the customer requirements to engineering specification is highly mature in their use to improve the customer relationship; suggestions given by the employees to improve the manufacturing process are highly mature in their use to involve people in improvement activities; adequate finance and marketing activities are highly mature in their use in terms of system (support core work); and effective deploying of the policy goals of the organization is highly used by the organizations in terms of leadership.

4.5 Input and Output Variables (CI Strategies and Performance Parameters)

In the present study, eight key CI strategies (A1, A2, A3, A4, A5, A6, A7, A8) including supplier development, process (just in time), people (total involvement), total quality management, system (support core work), leadership, total productive maintenance and customer relationship, and six manufacturing performance parameters (Z1, Z2, Z3, Z4, Z5, and Z6) including organization achievements, productivity, cost, quality, delivery, and safety have been taken as shown in Fig. 4.3.

4.5.1 Reliability Analysis of Input and Output Factor

Cronbach's alpha for various CI strategies and performance parameters has been evaluated to ascertain the reliability of the input and output data collected through the "CI Questionnaire." A cutoff loading of 0.45 has been used to screen out variables that are weak indicators of the construct. The Cronbach's alpha values for all the input and output categories, in excess of 0.7 (Singh & Khamba, 2010), indicate the significantly high reliability of data for various input and output categories. The values of Cronbach's alpha for various input and output categories have been shown in Table 4.4.

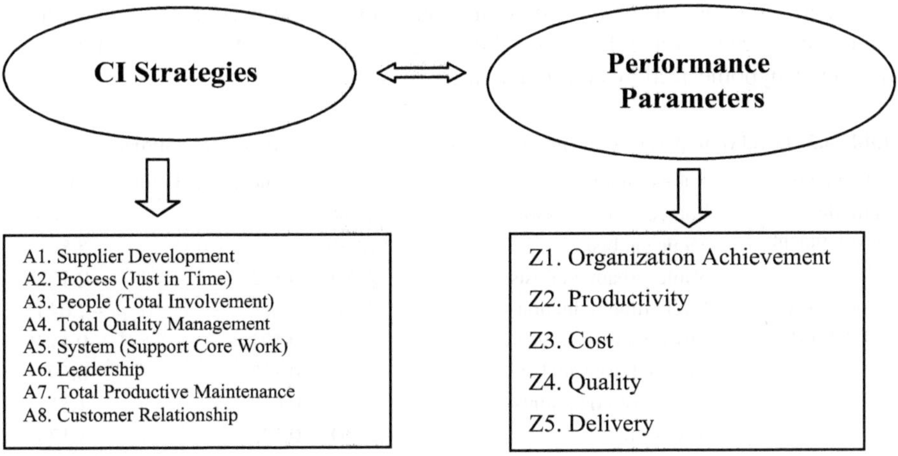

Fig. 4.3 Contribution of CI strategies in realizing performance parameters

Table 4.4 Reliability coefficients of various input and output parameters

Input parameters		Cronbach's alpha (α)
1	Supplier development	0.775
2	Process (just in time)	0.791
3	People (total involvement)	0.786
4	Total quality management	0.799
5	System (support core work)	0.728
6	Leadership	0.722
7	Total productive maintenance	0.755
8	Customer relationship	0.788
Output parameters		
1	Organization achievements	0.804
2	Productivity	0.714
3	Cost	0.786
4	Quality	0.760
5	Delivery	0.886
6	Safety	0.774

4.6 Level of Importance of Sub-strategies of CI Approach

In order to explore the level of importance of different CI strategies in terms of eight main strategies evolved in improving the manufacturing performance, z-distribution or z-test {for samples of size (n), more than 30, called large samples, z-distribution is applicable} has been applied. The level of importance of different CI sub-strategies has been measured from the values of mean and the significance level has been tested on the basis of z-statistic. The calculated z-statistic is compared with the critical values of z at the level of significance adopted (if the calculated value of z-statistics is less than table value of z at the level of significance adopted, it is concluded that sample data is significant). Table 4.5 shows the values of mean, standard deviation, z-statistics, and a hypothesized mean for sub-strategies of CI approach.

Table 4.5 Level of importance of sub-strategies of continuous improvement approach

CI strategy	Sub-strategy	Mean	Standard deviation	z-Statistics
Supplier development (A1)	Total cost management	2.158	0.393	−1.206[a]
	Value analysis	2.930	0.514	5.861[c]
	Value stream analysis	**2.801**	0.812	2.122[b]
Process (just in time) (A2)	Cycle time reduction	2.603	0.664	−2.044[b]
	Process flow analysis	2.772	0.691	−0.479[a]
	Material flow analysis	2.861	0.812	0.693[a]
	Theory of constraints	2.683	0.677	−1.812[a]
	Redesign	**2.891**	0.733	1.175[a]
	Workflow analysis	3.019	0.677	3.180[c]
People (total involvement) (A3)	Internal training and monitoring	2.198	0.490	−1.512[a]
	Manager development	2.881	0.604	2.829[c]
	Suggestion system	**2.950**	0.852	2.523[b]
	Team-based improvement	2.772	0.661	0.932[a]
	Principles of KAIZEN	2.752	0.740	0.564[a]
Total quality management (A4)	5S	2.653	0.727	−2.162[b]
	Error proofing analysis	**2.960**	0.598	2.080[b]
	Failure mode effect analysis	2.831	0.837	0.261[a]
	Six sigma	2.584	0.790	−2.468[b]
	Standardization	3.019	0.677	3.111[c]
System (support core work) (A5)	Finance	2.534	0.625	−2.380[b]
	Marketing	2.752	0.654	0.114[a]
	Process flow mapping	**2.861**	0.761	1.534[a]
	Support and administrative KAIZEN	2.831	0.762	1.141[a]
Leadership (A6)	Vision alignment and direction	2.445	0.574	−6.240[c]
	Policy deployment	2.960	0.662	2.404[b]
	Recognition	**3**	0.812	2.449[b]
Total productive maintenance (A7)	Preventive maintenance	2.415	0.570	−1.733[a]
	Autonomous maintenance	2.861	0.616	1.032[a]
	Equipment restoration	2.980	0.860	2.967[c]
	Change over reduction: SMED	2.772	0.760	−0.340[a]
	Minor stoppage elimination	**2.960**	0.564	2.491[b]
Customer relationship (A8)	Quality function deployment	2.485	0.593	−2.442[b]
	Customer quality, cost, and delivery analysis (QCD)	**3.009**	0.624	1.222[a]

Z critical (0.05) = 1.96
Z critical (0.01) = 2.58
[a]Significant at 5% level
[b]Significant at 1% level
[c]Not significant

4.6.1 Result Discussion of Level of Importance of Strategic Implementation of CI Approach

Table 4.5 depicts that the sample statistic (z-statistic) for the sub-strategies including value analysis, workflow analysis, manager development, standardization, vision alignment, and direction and equipment restoration lies in the rejection region which signifies that the respondent's claims regarding these strategies are not justified so these strategies are not included in the results. The results of investigation demonstrated that the process of directly observing the flow is highly significant in working continuously with the suppliers or value stream analysis is the most significant approach for supplier development, and managing the financial outcomes of activities encompassing all operations is least significant towards supplier development; redesigning the various aspects of product is highly significant for providing the product at right time, right place, right quantity, and right quality to the customer or redesigning is the highly significant approach of just-in-time manufacturing, and cycle time reduction is the least significant JIT technique for continuous improvement; suggestions from the employees at every level of organization play an important role in involving the people in continuous improvement activities or suggestion system plays a significant role in carrying out improvement activities effectively; internal training and monitoring are the least effective techniques to involve people in improvement activities; reducing human mistake by foolproofing is the most effective TQM technique for making small incremental improvements in the manufacturing system processes; six sigma is the least effective TQM approach for CI implementation; flowcharts of all types of activities are highly useful in terms of support core work for improvement purposes, and adequate finance is not so important supportive tool for small incremental improvements; giving rewards for most useful outcome is the significant leadership approach for carrying out improvement activities, and communicating the vision and clear goals of CI implementation is the least effective in terms of leadership; and customer quality, cost, and delivery analysis is the more important technique than quality function deployment for improving customer relationship or QCD analysis appreciably aimed at meeting the goals of continuous improvement through strong customer relationship.

4.7 Correlation Between Input and Output Variables (Validation of Hypotheses: H1)

On the basis of the responses received from the industry, an assessment of association of various CI strategies with key performance parameters has been made. In order to explore the role of CI strategic implementation towards manufacturing performance improvement, the Pearson correlations have been worked out to ascertain the significant factors contributing to success of the CI strategic implementation

Table 4.6 Correlation matrix between input and output parameters

	Z1	Z2	Z3	Z4	Z5	Z6
A1	0.0529[a]	0.0379[a]	−0.0004[a]	−0.2604[a]	0.1765[a]	0.0925[a]
A2	0.0031[a]	−0.0875[a]	0.1297[a]	−0.1097[a]	−0.0640[a]	0.1174[a]
A3	−0.0764[a]	0.0543[a]	0.0488[a]	0.0287[a]	0.0162[a]	0.0111[a]
A4	−0.1819	0.0764[a]	−0.0739[a]	0.2161[b]	0.1530[a]	−0.1166[a]
A5	−0.0289[a]	−0.1509[a]	−0.2212[a]	−0.2059[a]	−0.1227[a]	−0.0329[a]
A6	0.0054[a]	−0.0235[a]	−0.0269[a]	−0.0694[a]	−0.1227[a]	0.0424[a]
A7	0.0198[a]	0.0792[a]	0.0135[a]	0.0701[a]	0.0438[a]	0.0341[a]
A8	−0.2088[a]	−0.0009[a]	0.0773[a]	−0.2168[a]	0.0166[a]	−0.1783[a]
Overall	0.1132[a]	0.0102[a]	−0.0096[a]	0.1771[a]	0.0091[a]	−0.0221[a]

[a]Correlation is significant at 5% level (critical value = 0.1965)
[b]Correlation is significant at 1% level (critical value = 0.2550)

program in the organizations. The correlation analysis depicts how CI strategies are related to different benefits of CI or important benefits that these manufacturing organizations had achieved after successful implementation of CI strategies. The positive value of Karl Pearson coefficient of correlation depicts that each strategy has a positive effect on the manufacturing performance parameters of continuous improvement approach. The highest value of correlation coefficient and least value of negative correlation have been tested at the level of significance adopted (if the calculated value of correlation coefficient is less than table value at (n-2) degree of freedom, it is concluded that sample statistic is significant). Table 4.6 shows the values of Karl Pearson coefficient of correlation calculated for eight (08) main strategies and performance parameters of CI (correlation matrix between inputs and outputs of CI).

In order to investigate critical success factors, the significant correlations thus obtained as a result of Pearson's correlation are validated through multiple regression analysis, as depicted in Table 4.7. The notations depicted in the table include β = regression coefficient (beta coefficient) and R square = multiple correlation coefficient. The significant factors with beta (β) significance level, R-square values, and F-values for each performance parameter are indicated.

4.7.1 *Result Discussion of Correlation Between Input and Output Factors*

Table 4.6 depicts that all the highest positive and least negative coefficients are significant at 5% significance level and 1% significance level which indicates high statistical stability of input and output variables depicting manufacturing performance improvement. Moreover multiple regression implies that there is a strong contribution of various performance parameters as depicted in Table 4.7 with the respective CI strategies evolved in manufacturing system processes. The highest

Table 4.7 Multiple regression analysis between different variables

Dependent variables	Independent variables	Beta value	*t*-Value	*p*-Value	*R*-Square	*F*-Value
Z1	A1	0.0425	5.2236	0.0060	0.31	5.5746
	A7	0.0245	2.8836	0.0085		
Z2	A7	0.0786	5.6586	0.0057	0.91	4.4896
	A4	0.0598	5.2666	0.0059		
Z3	A2	0.1552	3.1623	0.0024	0.194	3.8106
	A8	0.0438	5.1651	0.0061		
Z4	A4	0.2879	2.8725	0.0408	0.466	2.3996
	A7	0.0043	2.5379	0.0097		
Z5	A1	0.2831	2.6845	0.0397	0.648	3.4001
	A4	0.3339	3.8794	0.0631		
Z6	A2	0.2076	2.9118	0.0364	0.169	5.8424
	A1	0.0981	3.5565	0.0579		

positive correlation coefficient is found between supplier development and delivery ($r = 0.1765$). Working continuously with suppliers or supplier development practices highly aimed at improving the delivery reliability of the product manufactured has a negative impact on the cost and quality-related aspects of the organization. The highest positive correlation is found between process (JIT) and cost ($r = 0.1297$), Just-in-time manufacturing most significantly aimed at reducing the cost associated with manufacturing system processes has a negative impact on productivity, quality, and delivery aspects; The highest positive correlation coefficient is found between people (total involvement) and productivity ($r = 0.0543$). Involvement of people in improvement activities significantly aimed at increasing productivity has a negative impact on organization achievements. The highest positive correlation coefficient is found between total quality management and quality ($r = 0.2161$). TQM approach significantly aimed at improving the quality of the product has a negative impact on organization achievements, cost-related issues, and safety aspects. The least negative correlation is found between system (support core work) and organization achievement ($r = -0.0289$). Supportive strategies have highly negative impact on organization achievements and also on other manufacturing performance parameters. The highest positive correlation is found between leadership and safety ($r = 0.0424$). Effective leadership significantly aimed at improving the safety aspects of the employees working on production equipment of the manufacturing organizations has a negative impact on productivity, cost, quality, and delivery aspects. The highest positive correlation is found between TPM and productivity ($r = 0.0792$); total productive maintenance strategies significantly aimed at improving productivity of the manufacturing organization also have a positive impact on all other performance parameters. The highest positive correlation is found between customer relationship and cost ($r = 0.0773$). Customer relationship plays a significant role in reducing the cost of the manufacturing system processes and has a negative impact on the organization achievements, productivity, quality, and safety aspects. Overall strategic implementation of CI approach significantly aimed at improving the

Table 4.8 Reliability coefficients of various input and output parameters

Input parameters		Cronbach's alpha (α)
1	CI driving values	0.742
2	CI enabling values	0.842
3	CI infusion values	0.757
4	Customer relationship	0.757
Output parameters		
1	Organization achievements	0.804
2	Productivity	0.714
3	Cost	0.786
4	Quality	0.760
5	Delivery	0.886
6	Safety	0.774

quality of product manufactured by the manufacturing SMEs has a negative impact on cost and safety-related issues of the manufacturing system processes. This quality improvement can be enhanced by adopting proper policies, such as continuity of operations, and addressing losses and wastages associated with the production system. This leads to enhanced quality of production, leading to significant reduction in process defects and customer warranty problems, and to improved customer order compliance.

The hypothesis which states that there exists an overall association between CI strategies and performance parameters is significantly supported but in Indian context there exists some negative impact of these strategies on performance parameters. Moreover, CI approach significantly aimed at improving the performance of the organizations.

4.8 Role of Key Enablers in Improving the Performance of SMEs

In order to explore the key enablers and sub-enablers in improving the performance of SMEs, the key enablers and sub-enablers have been identified from literature which includes CI driving values, CI enabling values, CI infusion values, and customer relationship. Cronbach's alpha for various CI enablers and manufacturing performance parameters has been evaluated to ascertain the reliability of the input and output data collected through the "CI Questionnaire." A cutoff loading of 0.45 has been used to screen out variables that are weak indicators of the construct. The reliability of all the constructs meets Nunnally's recommended standard (Cronbach's alpha ≥0.7) for early-stage research. The Cronbach's alpha values for all the input and output categories, in excess of 0.7, indicate the significantly high reliability of data for various input and output categories. The values of Cronbach's alpha for various input and output categories have been shown in Table 4.8.

Table 4.9 Results of *z*-test for four key enablers

Key enablers	Mean	Standard deviation	*z*-Statistics
CI driving values (X1)	2.798	0.743	−0.382[a]
CI enabling values (X2)	2.849	0.761	0.291[a]
CI infusion values (X3)	**2.851**	0.744	0.331[a]
Customer relationship (X4)	2.808	0.771	−0.239[a]

Z critical (0.05) = 1.96
[a]Significant at 5% level

4.9 Level of Importance of Enablers and Sub-enablers of CI Approach

In order to explore the level of importance of different enablers and sub-enablers evolved in improving the manufacturing performance ascertained from CI implementation, *z*-distribution or *z*-test {for samples of size (*n*), more than 30, called large samples, *z*-distribution is applicable} has been applied to ascertain the importance level of CI approach. The focus of this analysis is to explore the level of importance of four (04) key enablers and sub-enablers of CI approach. The level of importance of different CI enablers and sub-enablers has been measured from the values of mean and the significance level has been tested on the basis of *z*-statistic. The calculated *z*-statistic is compared with the critical values of *z* at the level of significance adopted (if the calculated value of *z*-statistics is less than table value of *z* at 1 and 5% significance level, it is concluded that sample data is significant). Tables 4.9 and 4.10 depict the values of mean, standard deviation, and *z*-statistic calculated for four key enablers and sub-enablers of CI approach.

4.9.1 Result Discussion of Level of Importance of Key Enablers and Sub-enablers of CI Approach

Result explicitness of Table 4.9 depicts that *z*-statistics for all the four key enablers lies in the acceptance region (if calculated $|z|$ value is less than critical value of *z* at the level of significance adopted, it is concluded that sample statistics lies in the acceptance region, i.e., claims are significant). The CI infusion value is the key enabler for successful implementation of CI approach followed by CI enabling values, customer relationship, customer relationship, and CI driving values. Table 4.10 depicts that *z* statistics for 90% sub-enablers lies in the acceptance region (10% includes empathy, structure support, and speed). The claims regarding these 03 (three) sub-enablers are not justified as a result these sub-enablers are not included in the results. Results indicated that responsibility is the most important strategy in terms of CI driving values for carrying out continuous improvement activities followed by respect. Trust in terms of CI enabling values followed by openness and cooperation and humbleness. Explicit behavior change is rated the most important

Table 4.10 Results of z-test for sub-enablers of CI approach

Sub-enablers	Mean	Standard deviation	z-Statistics
CI driving values			
Respect	2.257	0.439	−1.237[a]
Responsibility	**2.841**	0.561	0.768[a]
Empathy	3.297	0.781	6.407[c]
CI enabling values			
Humbleness	2.702	0.781	−1.877[a]
Trust	**2.911**	0.749	0.829[a]
Openness	2.891	0.761	0.556[a]
Cooperation	2.891	0.747	0.566[a]
CI infusion values			
Leadership	2.782	0.743	−0.937[a]
Structure support	2.633	0.643	−3.401[c]
Control system	2.871	0.783	0.254[a]
Explicit behavior change	**3.118**	0.725	1.705[a]
Customer relationship			
Efficiency	2.227	0.466	−1.251[a]
Quality	2.881	0.587	1.241[a]
Competence	2.961	0.904	1.686[a]
Flexibility	2.623	0.705	−2.334[b]
Innovation	2.851	0.766	0.562[a]
Speed	3.019	0.761	2.788[c]
Capacity	2.742	0.821	−0.808[a]
Uniqueness	2.941	0.718	1.846[a]
Reliability	**3.029**	0.842	2.538[a]

Z critical (0.05) = 1.96
Z critical (0.01) = 2.58
[a]Significant at 5% level
[b]Significant at 1% level
[c]Not significant

CI infusion value for incremental improvements in industry followed by control system and leadership; and reliability in terms of customer relationship plays a most important role in implementing continuous improvement approach followed by Competence, uniqueness, quality, innovation, capacity, flexibility, and efficiency.

4.10 Correlation Between CI Enablers and Manufacturing Performance Parameters (Validation of Hypotheses: H2)

On the basis of the responses received from the industry, an assessment of association of key enablers with manufacturing performance parameters has been made. In order to explore the role of implementation of key enablers towards manufacturing

Table 4.11 Correlation between input and output parameters

	Z1	Z2	Z3	Z4	Z5	Z6
X1	−0.215[a]	−0.106[a]	−0.068	−0.048[a]	−0.123[a]	−0.131[a]
X2	0.053[a]	0.102[a]	−0.054[a]	0.011[a]	0.002[a]	0.021[a]
X3	0.092[a]	−0.019[a]	0.018[a]	−0.139[a]	−0.018[a]	0.126[a]
X4	0.096[a]	0.063[a]	0.204[b]	0.066[a]	0.002[a]	0.186[a]
Overall	−0.008[a]	0.005[a]	0.009[a]	−0.127[a]	−0.082[a]	0.074[a]

[a]Correlation is significant at 5% level (critical value = 0.1965)
[b]Correlation is significant at 1% level (critical value = 0.2550)

Table 4.12 Multiple regression analysis between different variables

Dependent variables	Independent variables	Beta value	*t*-Value	*p*-Value	*R*-Square	*F*-Value
Z1	X4	0.1265	2.8434	0.0413	0.156	7.7727
	X3	0.6813	3.7924	0.0043		
Z2	X2	0.8761	2.9441	0.03474	0.1301	10.6459
	X4	0.7605	4.5024	0.00616		
Z3	X4	0.1476	2.8306	0.04081	0.7316	8.3611
	X3	0.2019	4.0185	0.00985		
Z4	X2	0.2399	3.2066	0.00836	0.488	7.2404
	X4	0.1294	4.6834	0.0495		
Z5	X2	0.4757	3.3021	0.0097	0.124	11.0005
	X4	0.5123	4.0199	0.00984		
Z6	X4	0.5243	2.7311	0.04656	0.45096	8.31406
	X3	0.1775	3.0247	0.0308		

performance improvement, the Pearson correlations have been worked out to ascertain the significant factors contributing to success of the CI enablers in the organizations. The correlation analysis depicts how different enablers are related to different benefits of CI or important benefits that these manufacturing organizations had achieved after successful implementation of these enablers. Table 4.11 shows the values of Karl Pearson coefficient of correlation calculated for four (04) key enablers and performance parameters of CI (correlation matrix between input and output parameters). In order to investigate critical success factors, the significant correlations thus obtained as a result of Pearson's correlation are validated through multiple regression analysis, as depicted in Table 4.12. The notations depicted in the table include β = regression coefficient (beta coefficient) and *R*-square = multiple correlation coefficient. The significant factors with beta multiple regression coefficient (β), *R*-square values, and *F*-values for each performance parameter are indicated.

4.10.1 Result Discussion of Correlation Between CI Enablers and Manufacturing Performance Parameters

The results of correlation analysis demonstrated that all correlation coefficients are significant at 5% level and 1% level. (All the values of correlation coefficients are less than critical values of correlation coefficient at the level of significance adopted.) The results of multiple regression show that various CI enablers and manufacturing performance indicators are closely associated, since the significance factor p works out to be less than 0.05 in all the cases and t-values are more than 2.62 (critical value of t). Further, results indicated that CI driving values have a negative effect on all manufacturing performance parameters. CI enabling values have a positive effect on all performance parameters but a negative effect only on cost-related issues. CI infusion values have a positive effect on organization achievements, cost, and safety aspects but also have a negative effect on productivity, quality, and delivery-related issues of the organization. Customer relationship shows a positive effect on all the manufacturing performance parameters. Overall performance of CI enablers shows a positive effect on productivity, cost, and safety and a negative effect on organization achievements, quality, and delivery-related performance parameters.

So, the hypothesis which states that CI enablers play a significant role in manufacturing performance improvement is not supported in Indian context as these enablers shows both positive and negative potential on performance parameters.

4.11 Benefits of CI Approach

On the basis of the responses received from the industry, an assessment of important benefits of various CI approach has been made. The importance level of different benefits has been calculated from the values of mean and the significance level has been tested from z-statistic. Mean, standard deviation, and z-statistics for different benefits of CI approach are determined in terms of organization achievements, productivity, cost, quality, delivery, and safety that have been shown in Table 4.13.

4.11.1 Result Discussion of the Findings for Important Benefits of CI Approach

Results indicated that z-statistic for the benefits "achieving high return on capital employed" and "enhanced production rates" lies in the rejection region, which depicts that these claims are not justified. So, these benefits are not included in the results. Further result explicitness depicts that the KAIZEN approach significantly aimed at improving the competitive image of the organization among different

Table 4.13 Results of z-test for different benefits

Benefits	Mean	Standard deviation	z-Statistics
Organization achievements			
Achieving high return on capital employed	2.059	0.311	−13.724[c]
Development of core-level competencies	2.504	0.756	0.284[a]
Increased profit	2.534	0.755	0.681[a]
Improved business performance	2.574	0.683	1.334[a]
Enhanced professionalism in the organization	2.603	0.679	1.781[a]
Improvement in competitive image of the organization	**2.623**	0.785	1.794[a]
Productivity			
Enhanced production rates	2.376	0.597	−4.107[c]
Improvement in overall equipment effectiveness	**2.762**	0.694	2.052[b]
Reduced material and operator movement	2.495	0.795	−1.584[a]
Effective process flow	2.693	0.612	1.192[a]
Reduced setup time	2.643	0.769	0.302[a]
Reduced cycle time	2.752	0.792	1.674[a]
Cost			
Reduced operating cost	2.485	0.687	−1.071[a]
Reduced cost of consumables	2.514	0.701	−0.624[a]
Reduced cost of overhead expenditure	**2.633**	0.845	0.894[a]
Reduced cost of inventories	2.594	0.763	0.468[a]
Reduced inspection cost	2.564	0.753	0.079[a]
Quality			
Reduced customer complaints and rejections	3.009	0.877	−2.401[b]
Effective utilization of quality tools	3.217	0.687	−0.024[a]
Reduced scrap and rework	3.297	0.686	1.135[a]
Reduced operator mistakes	**3.326**	0.776	1.389[a]
Reduced total percentage defectives	3.207	0.682	−0.17[a]
Improved quality of work-life	3.257	0.771	0.495[a]
Delivery			
Improved delivery reliability	2.405	0.532	−0.748[a]
Providing faster delivery of the product	2.445	0.842	0.119[a]
Customer satisfaction by providing at a time delivery	**2.485**	0.891	0.447[a]
Safety			
Reduced number of accidents	2.267	0.631	−0.473[a]
Maintaining health and safety standards	**2.326**	0.949	0.314[a]

Z critical (0.05) = 1.96
Z critical (0.01) = 2.58
[a]Significant at 5% level
[b]Significant at 1% level
[c]Not significant

Table 4.14 shows the classification of responses based on CI strategic implementation time period

S. No.	Phase	Characterization of organizations	No. of respondents
1	Introductory	Less than 4 years	21
2	Stabilization	Between 4–6 years	54
3	Maturity	More than 6 years	26

competitors in the market. The company's image in the consumer and business-to-business marketplace is critical for ultimate success in profits and relationships with the customer; improvement in overall equipment effectiveness of the various production equipment is highly significant for improving the productivity of the manufacturing system processes. KAIZEN methodology is highly effective in improving the equipment effectiveness by identifying different improvement aspects of the production machinery; cost of overhead expenditure is significantly reduced by applying KAIZEN methodology in the strategic manner, overhead factory-related costs that are incurred when a product is manufactured are significantly reduced by this approach; operator mistakes arising from the absenteeism of the employees leading to the poor quality of the product are removed by effective implementation of CI approach; quality defects arising from operator mistakes are significantly reduced by this approach; customer satisfaction by providing at a time delivery is the significant benefit in terms of delivery, which signifies that the methodology is highly helpful in providing the satisfaction to customer through effective delivery of the product by effecting improvements in equipment reliability, ensuring better upkeep of the production facilities; health and safety standards are significantly improved by effective deploying of various standards encompassing improving activities in the manufacturing system processes; health and safety standards are significantly maintained by effective implementation of KAIZEN methodology for the sake of improving the health- and safety-related aspects of the employees working in the manufacturing organization.

4.12 Validation of Hypothesis H3: Improvement in Manufacturing Performance Is a Function of Experience Gained by Manufacturing Organizations over an Extended Time Period

To investigate the effect of CI strategic implementation time period on the extent of manufacturing performance parameters accrued, the responses obtained from various manufacturing organizations have been classified into three categories, depending upon the experience gained over an extended period, as indicated in Table 4.14.

Table 4.15 summarizes the results of two-tailed "*t*" test employed to ascertain the improvements in various manufacturing performance parameters based on the CI strategic implementation time period. The table depicts the average and standard

Table 4.15 Two-tailed "*t*" test

Results of two-tailed "*t*" test for improvements in manufacturing performance parameters w.r.t. Time periods

Performance parameters	Mean and standard deviation						*t* (II/I) (*p*-value)	*t* (III/I) (*p*-value)	*t* (III/II) (*p*-value)
	Phase I		Phase II		Phase III				
	Introductory phase		Stability phase		Maturity phase				
	Less than 4 years		4–6 years		More than 6 years				
	$N = 21$		$N = 54$		$N = 26$				
	Mean	SD	Mean	SD	Mean	SD			
Z1	2.515	0.295	2.388	0.291	2.512	0.231	1.67 (0.1031)	1.58 (0.1237)	−0.04 (0.9655)
Z2	2.663	0.295	2.658	0.306	2.506	0.261	0.06 (0.9502)	−1.85 (0.0315)	−2.37 (0.0211)
Z3	2.503	0.364	2.547	0.403	2.607	0.317	−0.7 (0.0487)	0.36 (0.7216)	1.06 (0.0292)
Z4	3.246	0.345	3.349	0.323	3.012	0.442	−1.17 (0.0249)	−2.93 (0.0053)	−2.41 (0.0210)
Z5	2.388	0.533	2.492	0.512	2.435	0.386	−0.76 (0.4523)	0.4 (0.6883)	0.46 (0.0495)
Z6	2.25	0.676	2.2	0.529	2.519	0.685	0.3 (0.7676)	1.58 (0.0122)	1.77 (0.08510
Overall	2.594	0.231	2.641	0.198	2.598	0.172	−0.79 (0.0443)	−0.68 (0.0403)	0.1 (0.9194)

deviations, paired *t*-values {*t* (II/I), *t* (III/I), *t* (III/II)}, and significance values (*p*-values) of gains achieved for various manufacturing performance parameters. It has been observed that the mean values of the manufacturing performance improvements accrued in Phase II and Phase III are significantly higher than those obtained in Phase I, while the mean values of the manufacturing performance improvements accrued in Phase III are only marginally less than those obtained in Phase II. The overall effect of performance parameters signifies that there is marginal decline in improvement (not significant) in manufacturing performance parameters in the maturity phase (Phase III) as compared to stabilization phase (Phase II) that can be attributed to the fact that CI strategic implementation in the Indian industry is significantly beneficial for performance improvement in its maturity phase of implementation.

4.12.1 Result Discussion of Two-Tailed "T" Test

Table 4.15 depicts the comparative analysis of manufacturing performance improvements accrued in various time periods by effective CI strategic implementation program, using a two-tailed "*t*" test, at the significance level *p* of 0.05. The tests *t* (II/I), *t* (III/I), and *t* (III/II) have been conducted to ascertain the statistical difference in effectiveness of various manufacturing performance improvements accrued as a result of successful CI strategic implementation. The *t* (II/I) values and significance values of overall effect of performance parameters validate the significant realization of manufacturing performance improvements in the stabilization phase as compared to the introductory phase. Similarly, the *t* (III/I) values validate the significant realization of manufacturing performance improvements in the maturity phase as compared to the introductory phase. However, *t* (III/II) values reveal decline in improvement aspects (not significant) in the manufacturing performance parameters in the maturity phase as compared to the stabilization phase. Thus, the significance of manufacturing performance improvements accrued in Phase III as compared to Phase II could not be validated in the present context.

The present study shows that CI strategic implementation is not a typically matured incremental improvement process. It takes effective resource planning of manufacturing system processes, adequately assisted by organization culture through imbibing organizational improvement over a considerable period of time. As the findings of this empirical research are based on a survey conducted only in Northern India organizations, and the organizations selected for the study are in typical developing-country settings, the findings of this research work may not be readily extensible to other developed countries. Thus, in the context of the present study, it can be concluded that usually 4–6 years of whole-hearted CI strategic implementation is needed to realize significant results.

Table 4.16 z-Test applied to barriers in implementing CI approach

Barrier	Mean	Standard deviation	z-Statistic
Poor planning	2.534	0.671	−3.768[c]
Lack of management commitment	2.907	0.762	1.638[a]
Resistance of the workforce	2.792	0.863	0.064[a]
Lack of continuous training and education	2.861	0.787	0.954[a]
Teamwork complacency	2.712	0.840	−0.881[a]
Use of an off-the-shelf program	2.742	0.795	−0.555[a]
Failure to change organizational philosophy	2.891	0.859	1.223[a]
Lack of resources	2.683	0.733	−1.415[a]
Inability to change organizational culture	2.851	0.876	0.744[a]
Inability to build a learning organization that provides for continuous improvement	2.871	0.795	1.071[a]
Incompatible organizational structure and isolated individuals and departments	2.752	0.792	−0.431[a]
Inappropriate reward system	2.861	0.861	0.874[a]
Ineffective measurement techniques	2.931	0.851	1.701[a]
Paying inadequate attention to customers	2.712	0.711	−1.039[a]
Inadequate use of empowerment and teamwork	2.782	0.867	−0.051[a]
Poor interdepartmental relations	2.712	0.778	−0.951[a]
Lack of employee trust in senior management	2.881	0.778	1.222[a]
Politics and turf issues	2.722	0.826	−0.775[a]
Lack of strong motivation	2.633	0.808	−1.899[a]
Drive for short-term financial results	2.663	0.839	−1.473[a]
Lack of leadership	2.851	0.753	0.866[a]
Lack of time to devote to quality initiative	2.921	0.808	1.668[a]
Incompatible organizational structure and isolated individuals and departments	2.603	0.825	−2.223[b]
Use of a prepackaged program	2.792	0.852	0.065[a]
Inadequate knowledge or understanding of KAIZEN	**2.991**	0.806	2.537[b]

Z critical (0.05) = 1.96
Z critical (0.01) = 2.58
[a]Significant at 5% level
[b]Significant at 1% level
[c]Not significant

4.13 Important Barriers in Implementing CI Strategies

For analyzing the important barriers of KAIZEN, mean, standard deviation, and z-statistics have been calculated as shown in Table 4.16.

4.13.1 Result Discussion of Important Barriers in Implementing CI Strategies

Results indicated that the barrier "inadequate knowledge or understanding of KAIZEN" is the most important barrier in implementing CI strategies followed by "ineffective measurement techniques" and "lack of time to devote to quality initiatives." Results depict that "understanding of KAIZEN tools and technique" is the most significant barrier in implementing CI approach. Improper measurement techniques also affect the performance of manufacturing operations. Lack of time devotion to quality initiatives is the important factor that affects the performance of manufacturing SMEs. Management role in improvement activities plays a significant role in implementing CI approach. There is a need for proper training of proper application of KAIZEN tools and technique to ensure strategic success.

4.14 Concluding Remarks

Strategic implementation of continuous improvement approach is aimed at improving the quality of the product and improves the competitiveness of the SMEs in the wake of cut-throat competition. It has become evident that CI approach plays a significant role in achieving overall organizational performances for realizing the organizational objectives of sustainability and competitiveness. The continuous improvement approach is directly related to production; it can have profound effect on the cost of production. CI is increasing its importance by making small incremental improvements in the current manufacturing system processes and by achieving different types of benefits.

Chapter 5
Case Studies

5.1 Introduction

This chapter presents the case studies conducted at the following organizations: M/S Highway Industries Ltd., Ludhiana, and M/S Farm Parts Limited, Ludhiana, to analyze those areas and aspects, which have been reflected in the survey part.

5.2 Case Study I: Highways Industries Limited, Gill Chowk, Ludhiana

5.2.1 Introduction to Industry

Highway Company is one of the premier units of Hero Group of Companies. The company's office is situated in Punjab. The setup was established in the year 1971. The company is engaged in manufacturing and supply of auto parts to M/S Hero Moto Corporation Ltd. The wide range of products consists of crankshafts and kick lever for different range of motorcycles. The company is managed by a team of professionals with focus on:

- Customer satisfaction
- Continual improvement
- Prevention of pollution, hazards, and accident
- Compliance with legal and other applicable requirements

Beginning in 1971, the four decades of its eventful journey has established ABC Company a name to reckon with as reputed suppliers of hot-forged components and subassemblies to a number of industries vertical in two wheeler segments in India and abroad. With the vision of its founder—Umesh Munjal—the new dimensions

J. Singh, H. Singh, *Strategic Implementation of Continuous Improvement Approach*, SpringerBriefs in Operations Management,
https://doi.org/10.1007/978-3-319-93121-0_5

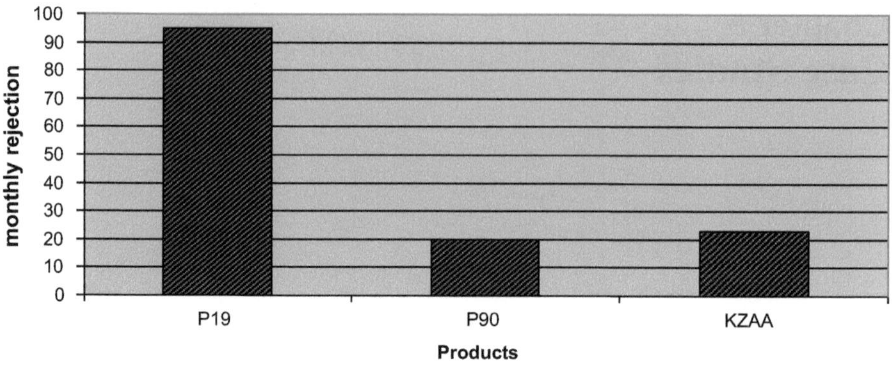

Fig. 5.1 Monthly rejection of crankshafts manufactured

have been added to the operations keeping excellence and quality as the benchmarks. Innovation, futuristic R&D, zero error, precision work, culture, and more all have added to elite performance of Highway. The different department includes forging shop, ultrasonic section, machine shop, and standard room.

Three KAIZEN improvement projects (KIPs) have been performed as discussed below:

5.3 KIP1: To Reduce Quality Rejection of Crankshaft P19 Using Six Sigma DMAIC Approach

5.3.1 Define Phase or Identification of Critical Areas

Project team of concerned persons has been constituted to define problem efficiently. The leader selected is a master black belt and team contains one black belt and three green belt members in it. A schedule chart for effective execution of respective DMAIC phases has been made by Six Sigma team. The rejections of various crankshafts (P19, P90, KZAA) coming from the customer have been analyzed. The crankshaft P19 contributes 70% of the total rejections of month. The monthly rejection for the various crankshaft manufactured by the company has been analyzed at different levels of production. The defective parts rejected are the sum of carrying in defects + bought-out semifinished defects + bought-out parts' defects. Figure 5.1 shows various quality rejections of different crankshafts manufactured. So, crankshaft P19 is considered a more critical product to carry out continuous improvement in its production processes.

Figure 5.2 shows the operation-wise rejection of the crankshaft P19 as determined at the operation level. 52 out of 95 rejections are caused due to ball height undersize at facing and centering operation. Furthermore this rejection analysis shows that common defects in facing and centering include ball height undersize, broken drill, double drill and center out, and total length undersize.

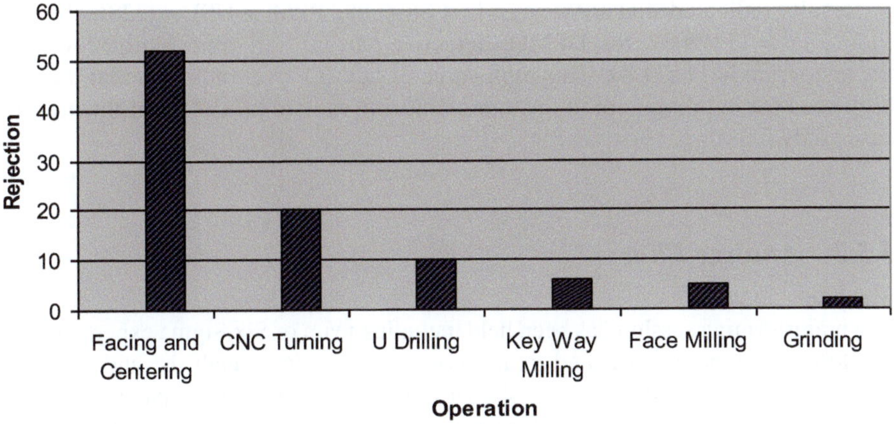

Fig. 5.2 Operation-wise rejection of the crankshaft P19

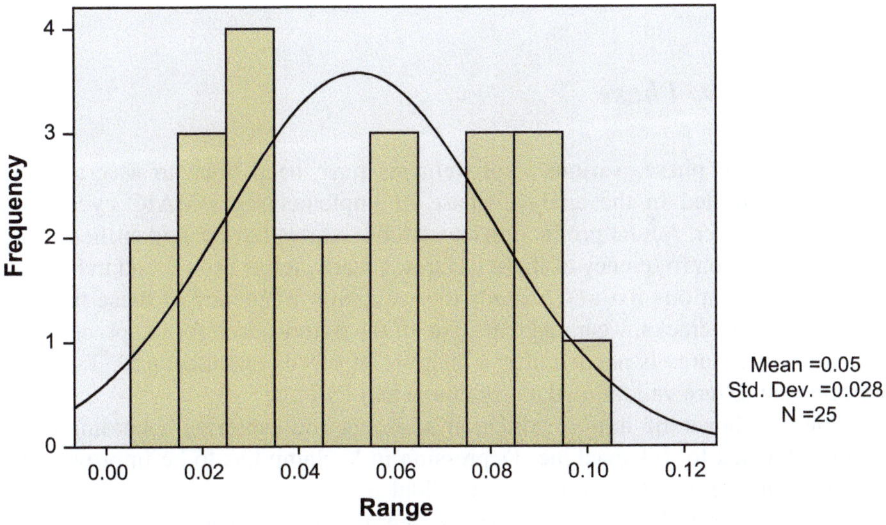

Fig. 5.3 Histogram of ball height undersize

5.3.2 *Measure Phase*

In order to verify the variation in the ball height ranging from 0.00 to 0.10 mm {LSL = 0.00; USL = 0.10}, the histogram has been prepared showing variation in the different readings of ball height undersize. The sample of 25 observations has been taken randomly and analyzed using SPSS software. Figure 5.3 indicates the different ranges of ball height at different frequency levels, which is helpful in depicting the rejection at million level.

Results indicated that PPM < LSL = 8892.98; PPM > USL = 2286.89; and PPM total = 11179.87. So, DPMO (defective part million opportunity) level for the histogram is 11179.87. This high value of DPMO level indicates that there is high need for improvement in the manufacturing system processes of the present case study.

5.3.3 *Analyze Phase*

The brainstorming session has been held including team of Six Sigma experts, HoDs of different departments, and managers. After critical analysis and expert's suggestions, fishbone diagram (as shown in Fig. 5.4) has been prepared depicting probable causes of ball height undersize at different levels of production. Root cause of the problem has been analyzed for taking appropriate preventive and corrective action through incremental improvements in their manufacturing system processes.

5.3.4 *Improve Phase*

In improvement phase, various improvements have been done to ascertain the problems identified in the analyze phase of implementing DMAIC cycle in a systematic manner. Actual production on various fixtures has been identified along with the validation frequency of these fixtures. Clearly, it has been found that actual production on various fixtures exceeds the validation frequency of these fixtures. This may lead to cracks, wear and tears, etc. of the fixtures used for the process. So, validation of fixtures is needed after a number of pieces manufactured. Table 5.1 indicates the fixture validation data for crankshaft P19 line.

Table 5.2 shows the fixture validation of facing and centering according to 04 fixtures attached to this machine. Depression in V clamp has to be improved and depression in pin-type locator has to be validated.

Table 5.3 shows the various problems identified or root cause of the problem (more flatness, new and unskilled labor, maximum die runs) and steps taken to improve the process.

5.3.5 *Control Phase*

To control the variation in the ball height, the SOP (standard operating procedure) and skilled matrix cum training need identification (providing on-job and off-job training as per recommendation of the different experts from various departments) and check sheets (to be filled by the operator daily) have been prepared to eliminate the roadblocks in facing and centering operation of the crankshaft P19.

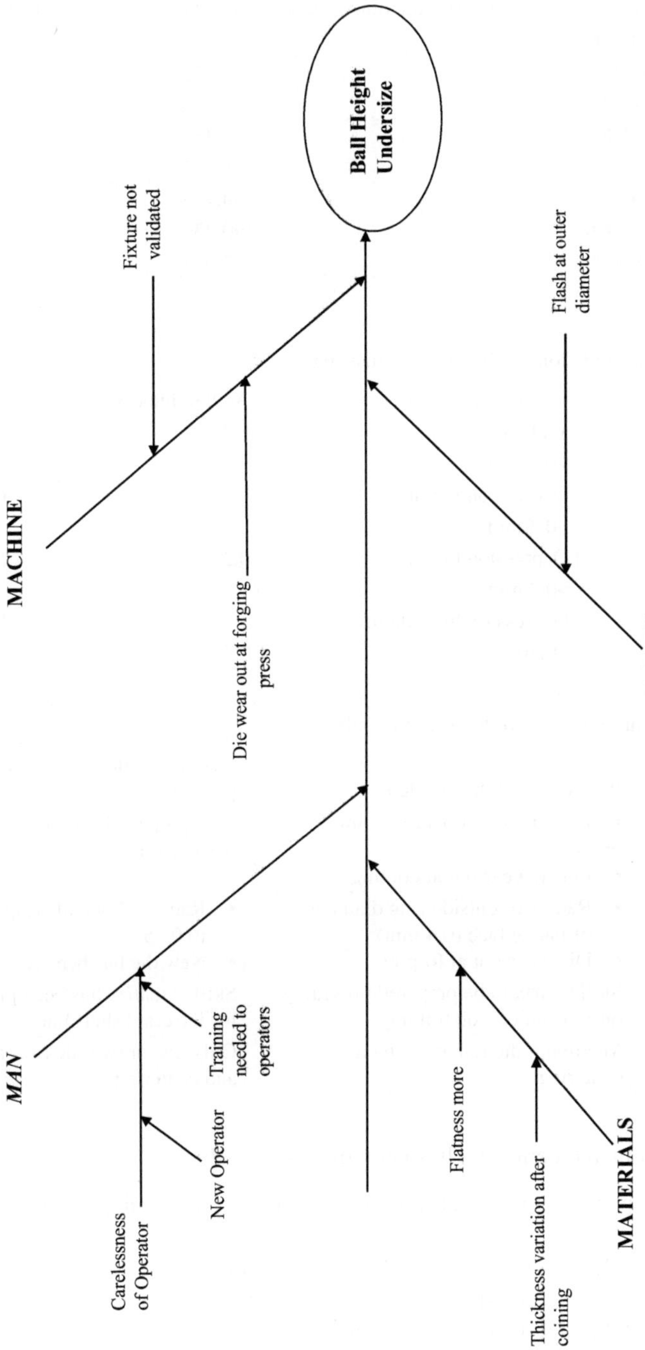

Fig. 5.4 Fishbone diagram depicting probable causes of ball height undersize

Table 5.1 Fixture validation data before implementing DMAIC cycle

Fixture no.	Validation frequency (production in pcs)	Actual production (production in pcs)
FC-CL-03	160,000	275,277
FC-CR-01	160,000	253,328
FC-CL-01	160,000	259,679
FC-CR-02	160,000	291,450
UD-CR-02	150,000	258,202
UD-CL-02	150,000	158,463
UD-CR-01	150,000	190,230
UD-CL-04	150,000	195,075

Table 5.2 Fixture validation for facing and centering

Fixture No.	Flatness specified	Actual flatness	Validation
FX-CS-0108	V clamp	0.34	Ok
	>0.02 mm		
FX-CS-0115	Depression in pin	0.40	Ok
	>0.3 mm		
FX-CS-0116	Depression in pin	0.25	Not ok
	>0.3 mm		
V BLOCK	Depression in V clamp		Not ok
	<0.30	0.5	

Table 5.3 Root cause analysis of total length undersize

Problem	Root cause of the problem	Activity made to improve the process
Flatness more at coining	• Pressure not proper at coining press • Punch wear-out at coining	New punch with proper pressure has been introduced
Flatness more at forging	• Radius of outside ring diameter of flange face (0.4 mm) • Die wear-out at forging	• Radius of outside ring reduced to 0.15 • New die has been introduced
New operator	Skill matrix to be prepared providing on-job and off-job training	Skilled matrix has been prepared for the crankshaft line
Die runs	Maximum die run at facing and centering	Only one or two die runs at facing and centering

SOP for facing and centering has been discussed below:

- Firstly, the operator will check the zero error of special gauge (ball height on surface plate).
- Operator will clean the machine and then fill the checklist.
- Keep the master piece of special gauges as per running component and remove the other model gauge from the machine.
- Whenever changing the setting, gauges should also be changed accordingly to the component.

Table 5.4 Check sheet for facing and centering

Check point	Day 1	Day 2	Day 3	Day 4
Cleaning the area				
Cleaning the workstation				
Lubrication of slide				
Coolant level as per level indicator				
Hydraulic pressure of clamping (15–20 kgf/cm^2)				
Guards				
Gauge color code				
Gauge identification				
Locator				
Resting pads				
Vertical and horizontal clamp				
Process operation sheet available				
Five-piece strand should be available				
Tool consumption report				

- Setting of tool by tool setter.
- After first five pieces, inspection by operator should be done.
- Operator will verify the length variation in center drill; if variation is found in length, return the drill to supervisor.
- Only one or two die runs should be at facing and centering.
- Maintain 5S around your machine (Table 5.4).

Figure 5.5 indicates that PPM < LSL = 10.22; PPM > USL = 18.47; and PPM total = 28.69. So, defective part million opportunity (DPMO) level for the histogram is reduced from 11179.87 to 28.69.

- PPM before Six Sigma = 11179.87
- Achieved PPM = 28.69
- Cost per piece = Rs. 110
- PPM reduction = 11151.18
- Monthly production = 120,000
- Saving per month = 11151.18 × 120000/10^6 = 1338.1416 pcs
- Yearly savings = 1338.1416 × 110 × 12 = Rs. 1766346.384

5.4 KIP2: To Reduce Setup Time Using Single-Minute Exchange of Die (SMED)

5.4.1 Methodology Adopted and Case Study Setting

The time associated with setup of forging press has been calculated and OEE associated with the forging press has been analyzed before and after implementing CI strategies. Different members of TPM departments of a manufacturing unit used

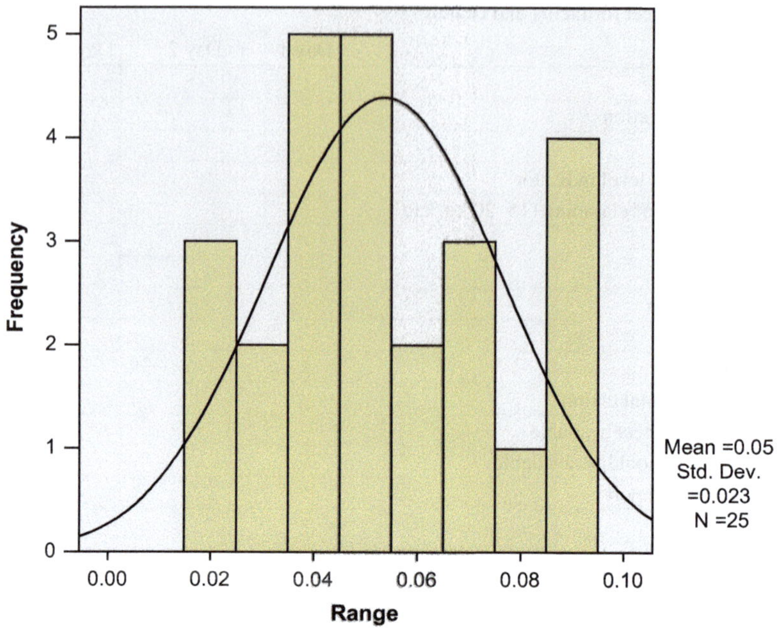

Fig. 5.5 Histogram after implementing Six Sigma DMAIC cycle

stopwatch technique to measure setup time of various activities associated with the progressive die setup (bottom and top blocker; bottom and top finisher) used in forging shop of the manufacturing unit. The setup time of internal activities (internal setup is that setup operation that can be done only when the machine is shut down) and external activities (external setup is that setup operation that can be done when the machine is still running) have been measured. These operations can be performed either before or after the machine is shut down. The time associated with setup of different dies used on the forging press has been calculated and OEE associated with this setup activities has been analyzed before and after implementing SMED in a systematic manner. The improvement in overall equipment effectiveness is tested for its significance by using statistical techniques to validate the data collection before and after implementing CI strategy, viz. SMED in the shop floor of the manufacturing industry under study. The total setup time associated with setup of die is 262 min. Table 5.5 shows all the activities associated with setup and time associated with these activities in setup of progressive dies.

The objective of SMED is to reduce the setup time. Setup time is the time elapsed at a work center from when the last good part of the current run is completed until the work center starts running the first good part of the next run. Long setup time resulted in a reduced number of setups, larger batch sizes, larger buffering work-in-process inventories, and poor process flow. In order to reduce the setup

Table 5.5 Activities associated with setup and time taken to perform these activities

Activities	Time taken (minutes)
Bolt-loosened bottom die	1
Bolt-loosened top die	3
Bottom die taken out	1
Top die taken out	1
Cleaning	1
Top finisher inserted	1
Top blocker inserted	1
Top die bolts tightened	3
Bottom finisher inserted	110
Bottom blocker inserted	90
Bottom bolt tightened	3
Die clamping bolt checked	1
Gap checked	1
Top die temperature	45
Bottom die temperature	
Total time	262

Table 5.6 Responsibility for SMED activities

Activities	Quantity	Responsibility
Arrange rods	4	Air man
Ensure that dies are properly heated in die-heating furnace	–	Die fitter Supervisor
Arrange spanner	02	Forger
Arrange Allen key	02	Forger
Pneumatic wrench	01	Forger
Remove bins from backside of forge press	–	Loader
Place billet bin of component to be forged near loader	–	Loader
Clamps for top die	01	Airman
Check all dimensions of top dies and bottom dies as mentioned in SMED check sheet		Die fitter
Cleaning of bottom holder grounded surface		Die fitter
Outside diameter of bottom and top dies (bulging effect)		Die fitter
Check temperature of top and bottom dies Top die = 175°C Bottom die = 220°C		Die fitter
Verify ejector pin length and diameter		Die fitter
Check tightness of bottom die and clamping bolts after heating		Die fitter

time associated with setup of progressive dies, the primary responsibilities have been assigned to air man, die fitter, forger, and loader. It results in reduction of setup time and is helpful in fluctuation of demand for different ranges of product. Table 5.6 shows the activities associated with setup of progressive dies and responsible persons for these activities.

5.4.2 OEE Improvement Using SMED

With the application of SMED technique, average setup time has been reduced.

$$\text{Downtime} = \text{Electrical breakdown} + \text{Die heating} + \text{Die change}$$
$$\text{Average setup time} = \text{Die heating} + \text{Die change} = 262\,\text{min}$$

$$\text{Availability} = \frac{\text{Loading time} - \text{Downtime}}{\text{Loading time}}$$

As the average setup time decreases, downtime also decreases. As a result, availability of production equipment increases as suggested by the above formula. Hence OEE increases considerably (Table 5.7).

5.4.3 OEE Improvement and Validation of Increase in OEE

Average OEE before implementing CI strategies = 61.41%

OEE after implementing CI strategies = 64.42%

Increase in OEE = 3.01%

This increase on OEE is validated by using paired sample t test

Null hypothesis:

HO1: There is no significant difference in the percentage improvement of OEE before and after SMED.

Alternate hypothesis:

Ha1: There is a significant difference in the percentage improvement of OEE before and after SMED.

Calculated *t*-value = 5.010902

The table value for 8 degrees of freedom at 5% significance is 2.306, which is less than 5.010902. Hence the null hypothesis is rejected. The result signifies that there is a significant difference in the percentage increase of OEE before and after implementing SMED. Hence, it is concluded that SMED implementation brings significant positive change to OEE improvement.

5.5 KIP3: To Reduce Forging Rejection of Crankshaft P19

5.5.1 Define Phase

The rejection for the month of January for different types of crankshafts including (P19 + P20 + KZAA) has been shown in Fig. 5.6. It is observed that maximum rejection is of crankshaft P19. The maximum rejection for P19 component is due to total length undersize as shown in Fig. 5.7. So, as a result this problem needs to be stabilized.

Table 5.7 OEE before implementing SMED

Day	Total time available (min)	Downtime (min)	Operating time (min)	Processed amount	Rejection	Availability	Performance efficiency	Quality rate	OEE (%)
3 March 2013	1320	490	830	6140	7	0.5	0.924	0.99	45.73
4 March	1320	545	775	5868	9	0.58	0.946	0.998	54.78
5 March	1320	435	895	5605	5	0.67	0.782	0.99	51.87
6 March	1320	530	990	8045	7	0.598	1.01	0.99	59.79
7 March	1320	375	965	7025	11	0.715	0.91	0.998	64.93
8 March	1320	325	915	6250	9	0.753	0.85	0.998	63.87
10 March	1320	304	925	6850	8	0.769	0.906	0.99	68.97
11 March	1320	345	920	6420	8	0.738	0.872	0.998	64.22
12 March	1320	405	890	6540	12	0.693	0.918	0.998	63.49
13 March	1320	275	910	7020	10	0.79	0.964	0.998	76.01

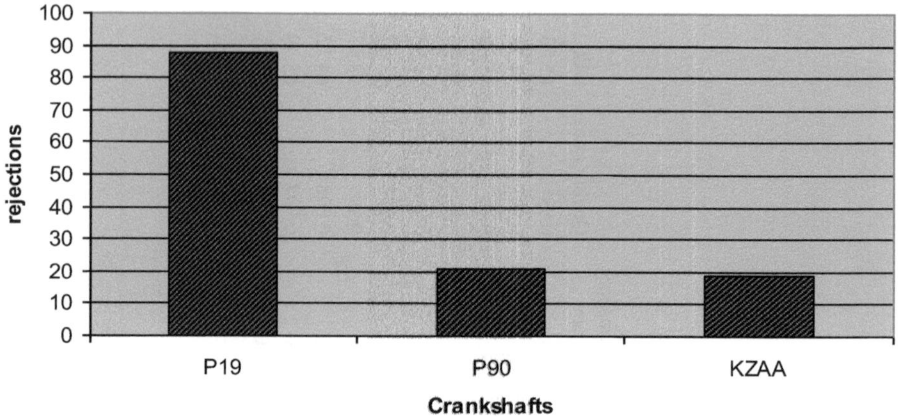

Fig. 5.6 Rejection of different crankshafts

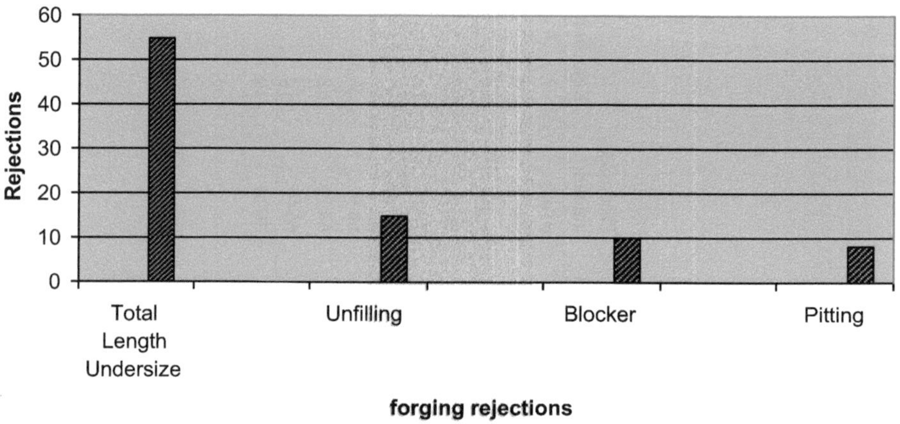

Fig. 5.7 Reason-wise rejection of crankshaft P19

5.5.2 *Measure Phase*

The histogram has been prepared with the help of SPSS software showing the mean and standard deviation of 20 samples taken randomly. Figure 5.8 shows the histogram of total length undersize having a standard length of 112.32.

Results indicate PPM < L.C.L = 0.03; PPM > U.C.L = 75655.22; and total PPM (out of limits) =75655.25.

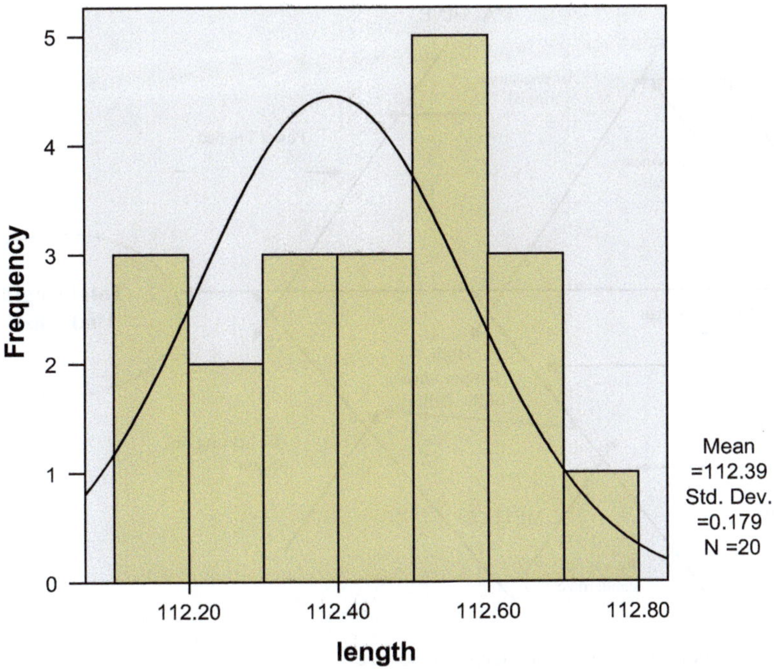

Fig. 5.8 Histogram of ball height undersize

5.5.3 Analyze Phase

Fishbone diagram has been prepared by panels of experts including production manager, quality manager, and maintenance manager showing the probable causes of total length undersize (Fig. 5.9).

5.5.3.1 Analyze Die Packing Plates

Die packing plate thickness is added to the die length for getting the required length of the crankshaft. From the supplier end, thickness of packing plates was not standardized, i.e., plates of various lengths were not standardized in various boxes. Table 5.8 shows the thickness of packing plates in different boxes.

5.5.3.2 Die Status for Forging of the Crankshaft P19

Total production at different stage has been analyzed and the actual production is more than the production standardized (around 5000). The total production exceeds the standardized production. So, continuous improvement is required. Table 5.9 shows production with dies and reasons to open.

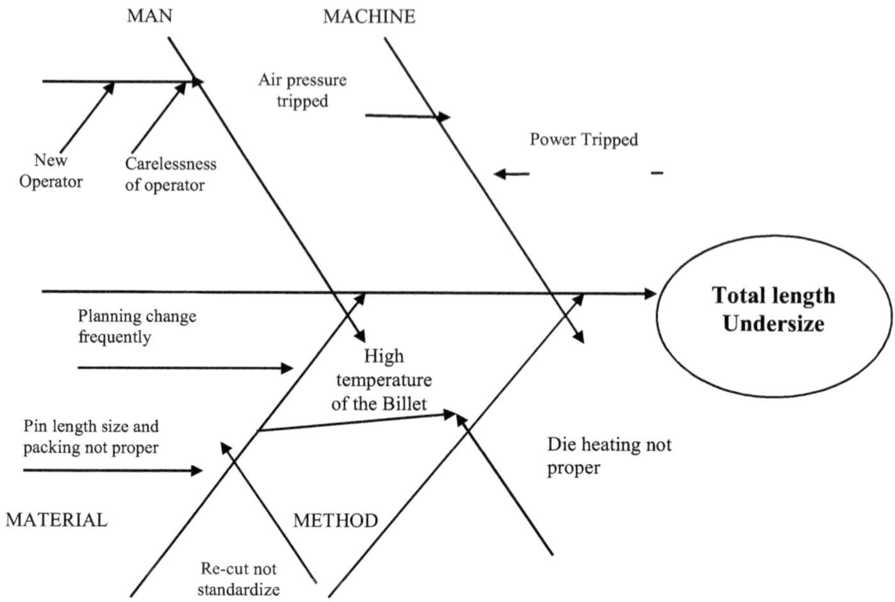

Fig. 5.9 Fishbone diagram of total length undersize

Table 5.8 Thickness of packing plates before standardization

Box No.	Thickness of packing plates (mm)
1	1–3
2	3–5
3	5–8
4	8–10
5	10–14
6	14–18
7	18–22
8	22–28
9	28–30

Table 5.9 Production in pieces and reasons to open die

Die status	Production in pcs	Reason to open die
New	5790	Worn out
Polished	1500	More flatness
First recut	3010	Worn out
Polished	3615	More flatness
Polished	2685	Crack
Second recut	4050	Planning changed
Total	20,650	Die continued

Table 5.10 Thickness of packing plates after standardization

Box No.	Packing plates (mm)
1	2.5
2	4
3	6
4	8.5
5	12
6	15
7	19
8	24
9	29

Table 5.11 Standardized total length of crankshaft P19

Die	Die no.	Length of die (mm)	Packing plate thickness (mm)	Total length required (mm)
Bottom finisher	BF-CR-1104	136	24	160
Bottom blocker	BB-CR-1102	136	24	160
Top blocker	TB-CS-1121	71	24	95
Top finisher	TF-CS-1110	89	6	95

5.5.4 Improve Phase

The thickness of the packing plates has been standardized in different boxes so as to obtain the required length of crankshaft P19 as shown in Table 5.10.

5.5.4.1 Standardize Total Length of the Dies

The standardized total length (length of die + thickness of packing plates) required is validated by using standardized plate thickness. The pin size length of bottom/ finisher and bottom blocker also got standardized as shown in Table 5.11.

5.5.4.2 Countermeasure for High Temperature of Billet (Table 5.12)

5.5.5 Control Phase

The standard operating procedure has been finalized to get the required length as shown below:

- Working area around forging press should be neat and clean.
- Tool shadow board should be placed at the proper area.

Table 5.12 Root cause analysis of total length undersize

Problem	Root cause of the problem	Activity made to improve the process
Tripping of workpiece in the induction coil	• Pump circulation is not proper • Input voltage not proper	• New pump has been introduced • Frequency of voltage stabilizer has been controlled
Improper temperature of furnace		

- All tools should be placed at their place (SMED table).
- Daily tools should be placed closer to forger.
- Fill the water spray tank and oil properly.
- Air gun should be in working condition.
- Die bolts and clamps should be tight.
- The length of bolts should be equal.
- Check the top die temperature and bottom die temperature.
- Ensure that billet heater temperature is 1210°, and then start production.
- Production should be started after first piece inspection by quality assurance person.
- Use the measuring instrument very carefully.
- Ensure that all gauges and instrument are calibrated.
- Use red bins for rejection, and yellow for rework.
- Use standardized pin and packing when setting a die; if any problem is found, inform the shift supervisor.
- Safety equipment must be used in forge shop.
- Maintain 5S in your working area.

5.5.5.1 Check Sheet for Forging Press

To enhance the control on total length, check sheet has been prepared, which contains detail of critical process parameters with their specified values and control procedure. The production supervisor has to fill this check sheet before start of improvement activities as shown in Table 5.13. The supervisor also has to check the activities after every 2 h during the regular shift of production by himself or herself.

5.6 Data Collection After Implementing DMAIC Approach

Histogram shows values of mean and standard deviation for a sample of 20 pieces randomly selected as shown in Fig. 5.10.

Results indicated that PPM $<$ L.C.L = 0.02; PPM $>$ U.C.L. =455.24; and PPM total (out of limits) =455.26.

Table 5.13 Check sheet for forging section

Description	Points to be checked	Day 1	Day 2	Day 3
Housekeeping	Area			
	Workstation			
Press	Power			
	Working switches			
	Counter meter			
	Line conveyor			
	Air pressure			
	Oil level			
Tools	Die conditions			
	Checkup gap			
	Uniform radial clearance			
	Availability			
Safety	Fume extractor			
	Emergency switch			

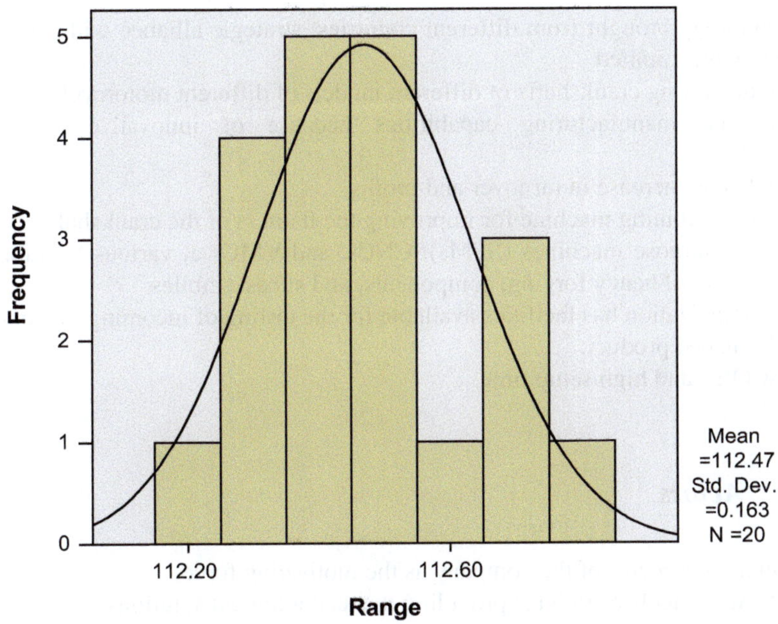

Fig. 5.10 Histogram showing data collection after improving total length undersize

5.7 Cost-Benefit Analysis

- Reduction in PPM = 75199.99
- Monthly production = 120,000
- Saving per month = 75199.99 × 120,000/10^6 = 1338.1416 pieces
- Yearly savings = 1338.1416 × 110 × 12 = 17. 65 lakhs

5.8 Flexible System Methodology

The flexible system methodology tries to resolve the end of continuum paradoxes, as it is based on a spectral paradigm, treating all of the system-based methodologies and techniques as lying on a continuum ranging from hard to soft and all problem situations also on a continuum ranging from unstructured to well structure. Hard and soft system-based techniques can be integrated in a number of ways. SAP-LAP (situation, actor, process-learning issues, actions suggested, performance expected) is a holistic framework that blends hard system and soft system paradigms (Sushil, 1999).

5.9 SAP-LAP Analysis for Case Study I

5.9.1 Situation

- Technology brought from different countries; strategic alliance with M/S Hero Moto Corp Limited.
- Manufacturing crankshafts of different models of different motorcycles.
- Enhanced manufacturing capabilities because of innovative efforts by everybody.
- Handsome increase in turnover and profits.
- A special coining machine for improving the flatness of the crankshafts.
- Special-purpose machines (SPMs), CNCs, and VMCs at various locations for machining of heavy forging, components, and subassemblies.
- The organization has facilities available for the testing of incoming raw material and finished product.
- Low OEE and high setup time.

5.9.2 Actors

- Managing director of the company as the motivating force.
- M/S Aarti Steels of India is providing the technological solutions.
- Top and middle management is dedicated a lot. Proper hierarchy is followed.
- Innovative engineers utilize the available strategies of CI approach by modifying the same to suit the local requirements.
- High-configuration workstations are used for improving activities continuously.
- Reputed auto parts manufacturing organization.

5.9.3 Process

- Keep upgradation of technology to enhance the production rates.
- Flexible workforce to tackle different types of CNC, VMC, and conventional machines.
- Enhanced manufacturing flexibility by launching new type of kick lever.
- Making use of local facilities, skills, and raw materials.
- Generating employment for a large number of persons.
- Significant reduction in setup and operation time through CI strategies, viz. SMED, 5S, Poka-Yoke, standardization, and redesign.

5.9.4 Learning Issues

- Strong technical support from M/S Vardhman Steels helps in grabbing the large market share.
- Higher production rates are the main objective of the organization.
- Company is investing regularly in machinery and equipment to keep it updated with improvements.
- Advanced tools like continuous improvement are used in most of the critical processes.
- Organization needs to change its policies and procedures to effectively monitor small incremental changes.

5.9.5 Actions Suggested

- Employees need to be updated with sophisticated CI tools.
- On-job and off-job training is needed to improve operator efficiency.
- Faculty from respective department should be involved in the training program.
- Considerable support is required from managers, supervisors, and workers for strategic implementation of CI approach.
- Make the technology more flexible by introducing volume flexibility, product flexibility, and process flexibility.
- Flexible staff hierarchy.
- Formation of cross-functional teams leads to improvement in processes.

5.9.6 Performance Expected

- Leading manufacturer of auto parts.
- Enhancement of market share.
- Strategic alliance resulting in increased export to other countries.
- Launch of new crankshaft in the Indian market.

5.10 Case Study II: Farm Parts Industries Ltd., Ludhiana

5.10.1 Introduction to the Industry and Company Strategy

The industry was founded in 1984 having more than 2700 employees. The plant is concerned with manufacturing and supply of forgings, components, and assemblies for three-point linkage for tractor farm equipment and Jondier tractor. The Farm Parts is equipped with a wide variety of CNC, VMC, and conventional machining equipment. The main aim of the company is to become global business and supply partner of core system to the off-highway market with a significant market presence in agricultural and construction market. Farm Parts has developed core competencies along the value chain like dual-shore manufacturing, warehousing solutions, and an unparalleled supply chain capability. With strategic location across the globe, Farm Parts with state-of-the-art equipment is able to survive its OEM customer in all main markets like Uniparts in India. The forging division of the company is equipped with state-of-the-art equipment to ensure world-class standards. The group has the capacity to forge 500 tons per month. Different ranges of product are manufactured like PTO shaft, draw and stray straps, weldons, balls, hub, hitch ball, and clutch sleeve. The company strategy is organic growth horizontal extension with existing customers by adding in new capabilities to address size ranges of existing products higher than current capabilities. The aim is to enhance design and R&D competence to become an "engineered solution provider" for global OEMs with offshore engineering and testing capability. Total quality is achieved through implementation of quality management systems. State-of-the-art testing and measuring equipment ensure perfect quality of all the products. Technologies like microscope, image analyzer, hardness tester, and surface roughness tester are used for ensuring the perfection and precision of our products.

5.11 X-Matrix for Policy Deployment

X-matrix is a document strategic plan of any organization that includes goals, strategies, strategic activities, and owners. It is also known as policy deployment (PD) document. The X-matrix has been prepared to deploy annual objectives, top-level improvement priorities, and area to be targeted by small incremental improvements. Deploying policy goals effectively in the organization is very helpful in achieving organizational goals. Different targets are set up which have to be followed on a daily basis. Primary and secondary responsibilities are assigned to the various employees of the organization to achieve policy goals. Figure 5.11 shows X-matrix for production department of organizations with targets to improve in various aspects (Fig. 5.12).

Tables 5.14, 5.15, 5.16, 5.17, and 5.18 show the various components of production, quality, and maintenance X-matrix indicating various objectives of

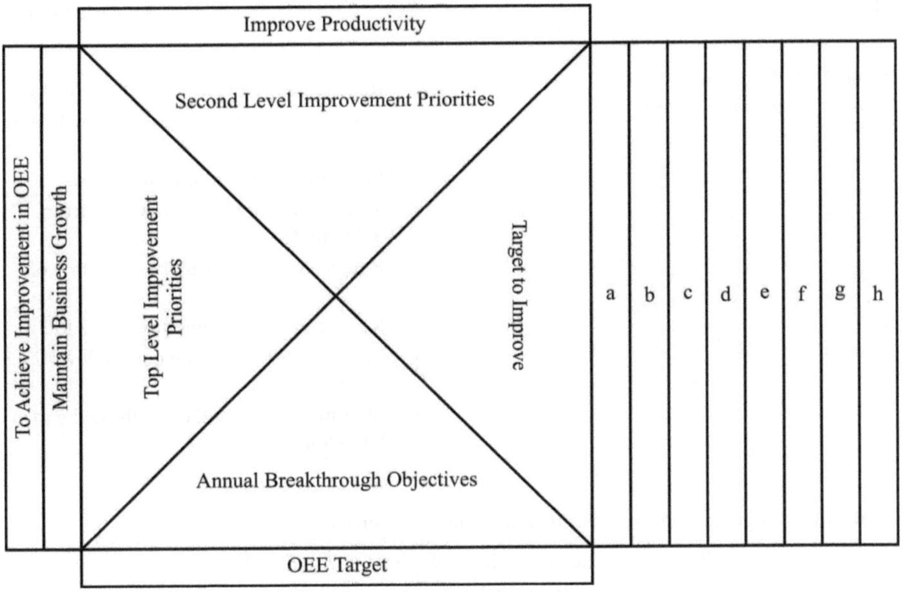

Fig. 5.11 Kenri's production X-matrix

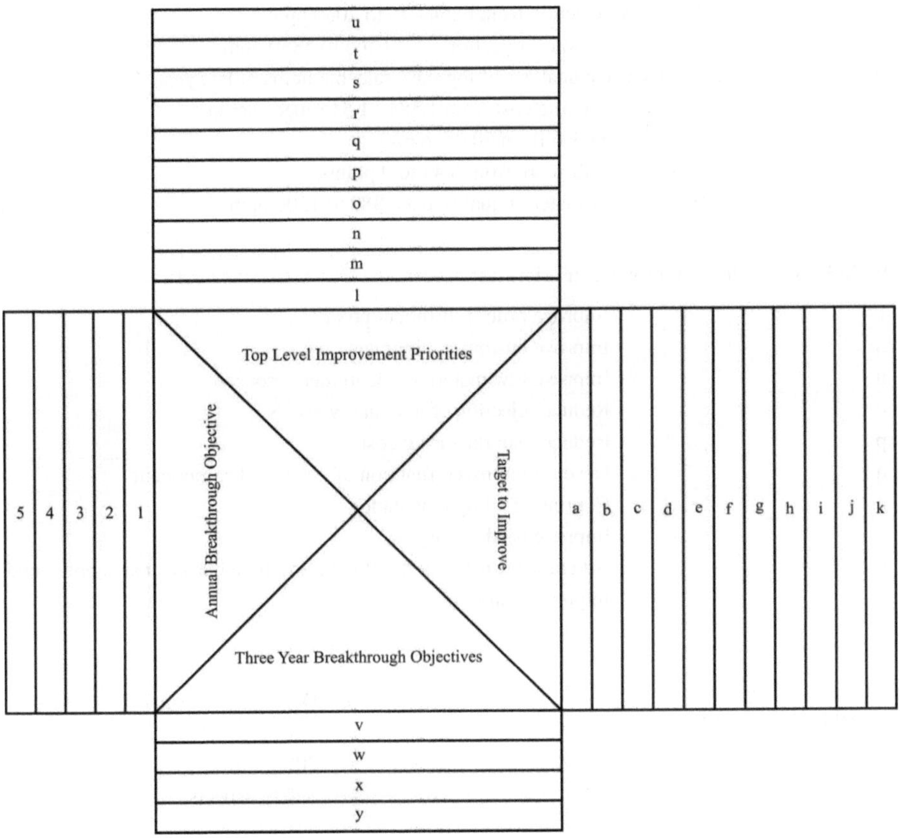

Fig. 5.12 Quality and maintenance X-matrix

Table 5.14 Target to improve areas in production

a	Reduce inefficiencies from 17.66 to 14%
b	Reduce machine breakdown from 4.13 to 3.75%
c	Setup time reduction from 4.13 to 2%
d	Reduce operator nonavailability from 3.755 to 2%
e	Raw material nonavailability from 1.65 to 1%
f	Reduce plan downtime from 10.7 to 3%
g	Improve 5S rating from current level 7.43 to 9 points
h	Start monitoring and controlling the man hours/ton

Table 5.15 Target to improve areas in quality and maintenance

a	To improve the time delivery from 93,702 to 85,000 ppm
b	To improve on-time availability of raw material from 124,875 to 15,000 ppm
c	To improve customer communication in sales from 96 to 99%
d	To improve customer communication in quality from 98 to 100%
e	To reduce rejection from 62644.75 to 2000 ppm
f	To reduce in-house rejection from 6164 to 5850 ppm
g	To make cost analysis of top 90% saleable items of last year
h	To reduce packing cost from 1.58 to 1.51 rupees per kg
i	To improve OEE from 50.75 to 63%
j	To improve 5S score from 8.44 to 9 points
k	Improve raw material quality from 880 to 1500 ppm

Table 5.16 Top-level improvement priorities

l	Improve order fulfillment process
m	Improve customer alignment
n	Improve raw material procurement process
o	Reduce rejection of ten quality goods
p	Reduce manufacturing cost
q	Develop improved function of machined component
r	Continue 5S implementation
s	Improve productivity
t	Add new business by developing machined product at supplier end
u	Improve quality process

Table 5.17 Three-year breakthrough objective

v	Delivery performance 5000 ppm
w	Quality level 200 ppm
x	20% Growth
y	Safety aspects

Table 5.18 Annual breakthrough objective

1	Delivery ppm to all customers 85,000 ppm
2	External quality ppm to all customers of 3500 ppm
3	Reduce product cost
4	Maintain 4.62% business growth
5	Accidents per 10,000 h worked

policy deployment action plan. The various activities are performed to deploy various targets as discussed below:

1. Identification of low OEE
    - To make yearly plan to train the CNC setter
    - To implement the training plan
    - To make monthly training plan for operator
    - To implement the production incentive
2. Reduce machine breakdown time
    - To prepare downtime data
    - To analyze and find the route cause
    - To prepare action plan
    - To implement reduced breakdown
    - To prepare the list of CNC machine
    - To replace old CNC machine
3. Reduce setup time
    - To make plan for SMED training
    - To ensure availability of tools, jig, fixtures, and drawing before setup
    - To reduce inspection time during setup
4. Nonavailability of operator
    - To ensure availability of worker
    - To maintain the base level
5. To reduce plan downtime
    - To prepare data for less order
    - To make plan to arrange order
    - To monitor monthly plan downtime
6. To reduce in-process troubleshooting
    - To prepare data and analysis for over-troubleshooting nature
    - To take action plan of the troubleshooting

Table 5.19 Achieved and targeted OEE

	April	May	June	July
OEE target	63%	63%	63%	63%
Achieved	51.16%	54.25%	54.36%	43.21%

Table 5.20 Plan and actual parts per millions (PPM)

Month	April	May	June	July
Plan ppm	2000	2000	2000	2000
Actual ppm	3397	2290	2859	2033

5.11.1 KIP 1: To Achieve 63% OEE

As shown in Table 5.19, average achievement in OEE is 50.75 which is less than the target of 63%. The route causes of low OEE are identified and countermeasures are taken to achieve the target. The causes are discussed below:

- Major breakdown of CNC-25 and CNC-2
- External and internal raw materials not available
- In-process troubleshooting

Countermeasures are discussed below:

- Maintenance department will provide the action plan to minimize the breakdown.
- SMED training provided to setter to reduce the setup time as per plan.
- Production planning and control department will ensure the availability of material as per plan.
- Identify the major reason of troubleshooting and provide the training to setter and operator.

5.11.2 KIP2: To Reduce Rejections to 2000 ppm

As shown in Table 5.20, average rejection in ppm is 2644.75 which is more than 2000, so action needs to be taken.

The causes are discussed below:

- Major rejection of wrong clamping due to use of other section manpower
- Major rejection of fixture fault of slip end, hitch ball, and actuation cover
- Major rejection of setting part

Countermeasures are discussed below:

- Identify the operators to provide the training of wrong clamping.
- Provide the training to operator about proper clamping of component on machine.

Table 5.21 Plan breakdown and actual breakdown

	April	May	June	July	August
Plan breakdown	1.39%	1.39%	1.39%	1.39%	1.39%
Actual breakdown	1.69%	2.03%	2.48%	1.47%	1.59%
Breakdown hours	574	1760	2069	1913	1679

- Modify and rectification of fixtures of rejection due to fixture fault.
- Identify the setter and reason of setting rejection.
- Provide the training to setter about first piece ok in setting.

5.11.3 KIP3: To Improve the Breakdown Hours (Table 5.21)

Major causes of breakdown of different machines:

1. CNC Machine

   - In CNC machine spindle head was defective.
   - In DLS section CNC-7 motor encoder cable jack was defective.
   - Spindle belt was broken due to defective motor pulley.
   - CNC-23 spindle encoder jack was loose.
   - CNC-9 tool clamp and unclamp lock was broken and carousel plate was damaged.

2. Milling Machine

   - M-31 motor belt was loose, coil was defective after tightening the motor belt, and hydraulic coil was replaced.
   - M-19 main gear box was damaged after gear sent to the market for hardening teeth cutting.
   - M-24 main gear box and bearing were damaged.
   - M-13 feed gear box clutch was broken and spindle shaft bush was loose.

3. Broaching Machine

   - Broach machine B-6 hydraulic cylinder seal was defective.
   - In U drill BS2 main gear was broken.

4. Drill Machine

Drill machine D-2, D-17, D-18 main gear shaft, and gear were broken.
Countermeasures are discussed below:

- Demand was raised for z-axis new encoder cable.
- Keep spindle motor pulley in spare stock.
- CNC machine motor encoder jack should be checked for preventive maintenance.

Table 5.22 MTBF of critical machine before preventive maintenance

Machine	Total planned hours	Total breakdown hours	No. of breakdowns	MTBF hours
CNC-12	572	4	3	189.33
VMC-4	572	17	4	186.33
Broach machine	572	28	2	272
Drill machine	572	13	3	279.5

- Clamp and unclamp lock should be replaced and keep carousel plate in spare stock.
- Motor belt and hydraulic coil should be checked for preventive maintenance.
- Keep M-19 gear in spare stock.
- Keep M-24 gear and bearing in spare stock.
- Keep M-13 gear and bearing in spare stock.
- Keep M-13 feed gear box clutch plate in spare stock.
- Keep B6 hydraulic cylinder seal in spare stock.
- Keep B92 main gear in spare stock.

- Drill D2, D17, and D18 main gear shaft and gear required in spare stock.

With this countermeasure, the target of plan breakdown has been achieved.

5.11.4 *KIP4: To Increase MTBF of Various Machines*

The critical machines have been identified, mean time between failures (MTBF) is calculated, and preventive maintenance schedule has been prepared to improve MTBF. The monthly, half-yearly, and annual maintenance schedule has been prepared. Table 5.22 shows MTBF of critical machines.

$$\text{MTBF} = -\frac{\text{Total production hours} - \text{Total breakdown hours}}{\text{Number of breakdowns}}$$

5.11.4.1 Preventive Maintenance Schedule for CNC-12 Machine

1. Monthly routine by maintenance

- Clean the coolant pipe, joints, filter, and tank.
- Check the hydraulic and pressure gauges for healthy operation.
- Check for any leakage in all distributed hydraulic and lubrication pipes.
- Clean the spindle belt encoder belt and turret belt.
- Inspect ball screw, bearings, and slides visually.

- Check main electric wires for proper tightness.
- Clean all fans and air filters.

2. Half-yearly routine by maintenance

   - Repeat all monthly checks.
   - Check the sliding guards for any play.
   - Check for any abnormal sound during running of machine.
   - Measure the neutral to earth voltage that should be below 0.3 to 3 V AC.

3. Annual routine by maintenance

   - Check hydraulic and lubricating oil filters. If require replace it.
   - Remove all side covers of machine and clean the machine fully from dust and chip.
   - Check all limit switches for healthy operation.

5.11.4.2 Preventive Maintenance Schedule for VMC-4 Machine

1. Monthly routine by maintenance

   - Clean all coolant filter and tank.
   - Check for any leakage in all distributed hydraulic and lubricating pipes.
   - Check the hydraulic pressure gauge for proper tightness.
   - Check main electrical wires for proper tightness.
   - Clean the compressor air filter.
   - Measure all safe input and output voltage.
   - Clean all fans and air filters.
   - Clean the automatic tool changer (ATC) and provide proper greasing to all wearing parts.
   - Check the lubrication oil level and clean the oil filter.
   - Open the top cover of *z*-axis spindle unit and clean all parts carefully.

2. Half-yearly routine by maintenance

   - Repeat all monthly checks.
   - Check the sliding guards and wipers for any play.
   - Measure the neutral to earth voltage that should be below 0.3–3 V AC.
   - During working of the machine check for any vibrations.

3. Annual routine by maintenance

   - Repeat all half-yearly checks.
   - Remove all side covers of the machine and clean the machine fully from dust and chips.
   - Check all limit switches for healthy operation.
   - Check the level of the machine using micron-level meter.

5.11.4.3 Preventive Maintenance Schedule for Broach Machine

1. Monthly routine by maintenance
   - Clean the hydraulic valve and pilot valve.
   - Clean the hydraulic oil filter, and replace it if required.
   - Check the movement and alignment of the piston.
   - Check the operation of hydraulic pump.
   - Check the limit switch condition.
   - Check the coolant pump and its leakage.
   - Check the motor load current equal to 20.25 Amperes.
   - Check the electrical connection, and tighten if required.
2. Half-yearly routine by maintenance
   - Repeat all monthly routine by maintenance.
   - Check the alignment of slide with reference to bed.
3. Annual routine by maintenance
   - Repeat all half-yearly checks.
   - Check the seal of the piston, and replace it if required.
   - Filter the hydraulic oil if required.

5.11.4.4 Preventive Maintenance Schedule for Drill Machine

1. Monthly routine by maintenance
   - Check the shaft of the machine for any outness.
   - Check the feed and feed gear for any wear and tear.
   - Clean the hydraulic filter and valve.
   - Check the tightness of electric connection and clean whenever required.
   - Check the motor load current.
   - Check the tightness of gear box.
   - Rectify all leakages found during check or intimated by operator.
2. Half-yearly routine by maintenance
   - Repeat all monthly checks.
   - Check the shaft and motor bearings, lubricate with grease, and if required change it.
   - Check all oil seals and packing; if required, change it.
3. Annual routine by maintenance
   - Repeat all half-yearly checks.
   - Completely open the gear box, and check the gears and bearings.
   - Check the level and condition of hydraulic oil (Table 5.23).

Table 5.23 MTBF after implementing preventive maintenance

Machine	Total planned hours	Total breakdown (hrs)	No. of breakdowns	MTBF (hours)
CNC-12	572	2	1	570
VMC-4	572	8	2	282
Broach machine	572	14	1	558
Drill machine	572	7	2	282.5

5.11.5 Increase in MTBF and Validation of Increase in MTBF

So average increase in MTBF from 189.29 to 358.125, i.e., about 168.84 h.

Null hypothesis:

HO3: There is no significant difference in the percentage improvement of MTBF before and after preventive maintenance.

Alternate hypothesis:

Ha3: There is a significant difference in the percentage improvement of OEE before and after preventive maintenance.

Calculated t = 2.776445.

Table value is 2.447 at 5% level, which is less than calculated *t*. As a result, null hypothesis is rejected. So, it is concluded that there is a significant increase in MTBF before and after preventive maintenance.

5.11.6 KIP5: To Increase Operator Efficiency by Technical Training

In order to increase the operator efficiency, technical training about CI strategies has been imparted to every employee of the organization. KAIZEN team will impart training on various principles of implementing the CI strategies. These strategies are used to reduce seven types of waste from the shop floor. Various principles behind these strategies are explained and followed by every employee in the organization. The main goal is to create a better process using small incremental improvements on a continual basis. Table 5.24 shows the production technical training chart for employees in the present study.

With the application of this production technical training, overall operator efficiency has been increased from 73.04 to 84.79 as shown in Table 5.25.

$$\text{Overall operator efficiency} = \frac{\text{Total pieces made by an operator} \times \text{SAM of operation}}{\text{Total hours worked} \times 60\,\text{min}}$$

Table 5.24 Production technical training chart

Topic	April	May	June	July	Aug.	Sept.	Oct.	Nov.	Dec.
SMED	√	√	√	√					√
Troubleshooting			√						√
X-matrix	√								
Quality control	√			√	√		√	√	√
5S		√				√		√	
Basic operating skills			√			√		√	
Autonomous maintenance		√				√			
Poka-Yoke								√	√

Table 5.25 Overall operator efficiency

	April	May	June	July	Aug.	Sept.	Oct.	Nov.	Dec.
OEE	73.04	75.41	75.75	80.66	82.14	82.07	84.44	83.51	84.79

5.12 Maintenance Job Card

Maintenance job card described the card that normally contains information of a job that is to be done in a certain maintenance facility. It contains the machine name, nature of fault, expected target for completion, scheduling, and any other additional instruction. It is usually used to authorize production people in a certain production plant. It is being filled by maintenance manager to check the nature of fault in the process or machine and cause of this fault.

Department---------------------- Machine/Equipment Name------------------------	Date--------------------
Nature of fault	Maintenance remarks
Section incharge sign----------------	Expected target for completion
Maintenance work satisfactorily completed Date:----------------- Time:------------------------ Section incharge----------------------	Maintenance inchrage:

5.13 Breakup of OEE and Rejection

5.13.1 Cause-Wise Breakup for OEE: The Percentage Breakup of CNC Machine Is Given Below

- Actual usage = 54.27%
- Planned downtime = 0.04%
- Low air pressure = 0.12%
- Tools/insert change = 0.19%

- Operator not available = 0.26%
- Inspection = 0.60%
- Internal raw material not available –0.60%
- External raw material not available –0.65%
- Preventive maintenance = 0.82%
- In-process troubleshooting = 1.94%
- Setting = 2.43%
- Breakdown = 3.68%
- Lunch/tea break = 8.24%
- Operator inefficiency = 9.41%
- Booking for samples = 16.75%

5.13.2 Cause-Wise Breakage of Rejection

- Power = 0.32%
- Machine maintenance = 0.70%
- Trial = 0.82%
- Handling = 4.12%
- Sample development = 5.76%
- Material fault = 13.49%
- Wrong clamping = 16.02%
- Outsource item = 16.72%
- Offset fault = 17.99%
- Forging fault = 19%

5.14 SAP-LAP on Case Study III

5.14.1 Situation

- Technology has been built in the organization.
- Leading manufacturer in different components of tractor parts.
- Enhanced manufacturing capabilities because of innovative efforts by every individual.
- Relative increase in turnover and profits even during recession period.
- Special-purpose machines (SPMs) and vertical machining centers (VMCs) at different sections for improving productivity.
- Modern facilities available in forging shop.

5.14.2 Actors

- Management of the company is providing the motivating force.
- M/S SKJ Limited is providing the technological support.
- Proper hierarchy has been followed.

- Innovative engineers utilize the available technologies by modifying the same to suit the local requirements.
- High-configuration workstations are used for design activities.
- Reputed tractor part manufacturing organization.

5.14.3 *Process*

- Keep upgradation of incremental technology to enhance the production.
- Flexible workforce to tackle different types of machines.
- Enhanced product flexibility by launching new parts of tractors.
- Making use of local facilities, operator skills, and raw materials.
- Generating employment for a large number of persons.
- Significant reduction in rejection rate through CI approach.

5.14.4 *Learning Issues*

- Strong technical support from M/S JONDIER tractors helps in grabbing the large market share.
- Higher productivity is the main objective of the organization.
- Company is investing regularly on machinery and equipment to keep it updated with new technologies.
- New technologies to be used in most of the critical processes.
- Organization needs to change its infrastructure to effectively utilize advanced techniques.

5.14.5 *Actions Suggested*

- Employees need to be updated with latest technologies.
- In-house training needs to be extended.
- Faculty from respective department should be involved in the training program.
- Considerable support is required from managers, supervisors, and workers for optimum utilization of technologies.
- Make the technology utilization more flexible by introducing volume flexibility, product flexibility, and process flexibility.
- Flexible motivation by management.
- Formation of cross-functional teams leads to new technology utilization.

5.14.6 *Performance Expected*

- Leading manufacturer of tractor parts
- Enhancement of market share

Table 5.26 Comparison between case studies

Point of difference	Highways Industries Ltd.	Farm Parts Industries Limited
Company strategy	To satisfy the need of customers with better quality of product	Zero-defect approach and muda elimination
Performance indicators	Trust, quality, and commitment	Superior quality product
Product mix	Crankshafts, kick levers, etc.	PTO shaft, draw and stray straps, weldons, balls, hub, hitch ball, clutch sleeve, etc.
SWOT	Strength: Personal attention to the customers; average growth of 40%; untiring dedication and leadership; highly qualified team of engineers; in-house R&D capabilities Weakness: Facing tough competition from other auto parts manufacturers. Opportunities: Fully integrated and state-of-the-art assembly facilitation for producing world-class tractors; R&D is a complete department in itself starting from designing up to development and implementation Threats: Not doing well in the automobile sector	Strength: Well-trained labor; marketing and operation teams are constantly creating new and effective strategies using modern management techniques Weakness: Not facing much competition Opportunities: A technology tie-up with major leaders in the world Threats: Sometimes laxity is made by workers
Relative performance of technology	Strategic implementation of CI approach	To have mastered the art of the KAIZEN principles
Benefits achieved	• Reducing rejection • Setup time reduction	• Increase in OEE • Improving overall operator efficiency • Increase in MTBF • Improvement in the breakdown hours
Strategies applied	• DMAIC • SMED	• Preventive maintenance • Policy deployment • Internal training and monitoring • Standardization

- Strategic alliance resulting in increased export to other countries
- Launch of new air-conditioned tractor parts in the Indian market

5.15 Comparison of Case Studies

Application of CI strategies is emerging as a strategic weapon in this highly competitive world. For this purpose, all the firms are implementing such strategies to adopt the organizations with the potential to gain earlier entrance to the market and offer higher quality products with improved consistency and reliability but have not so far successfully implemented for competitive success. Table 5.26 shows the comparison of industry performance achievements in three organizations.

Chapter 6
Synthesizing of Data

6.1 Introduction

In this chapter two-phase methodology using AHP and VIKOR has been applied to rank the different strategies of CI approach as per questionnaire filled by five respondents. It is an attempt to access importance level of CI strategic implementation in an organizational system to provide a framework in the selection of competitive advantages under uncertainty for strategic success. In the present study, three-level hierarchy model has been used to find out its measure of success possibility for CI strategic implementation as shown in Fig. 6.1.

6.2 Methodology Adopted

Figure 6.2 shows the flowchart for two-phase methodology applied.

6.2.1 Step 6.2.1: Degree of Preference

The degree of preference or intensity of the decision-maker in the choice of each pair-wise comparison used in this model is quantified on a scale of 1–9. This scaling process can then be translated in priority weight (scores) for comparison of alternatives. Even number (2, 4, 6, 8) can be used to represent compromises among the preference above. The suggest numbers used in this model to express degree of preference are shown in Table 6.1.

J. Singh, H. Singh, *Strategic Implementation of Continuous Improvement Approach*, SpringerBriefs in Operations Management,
https://doi.org/10.1007/978-3-319-93121-0_6

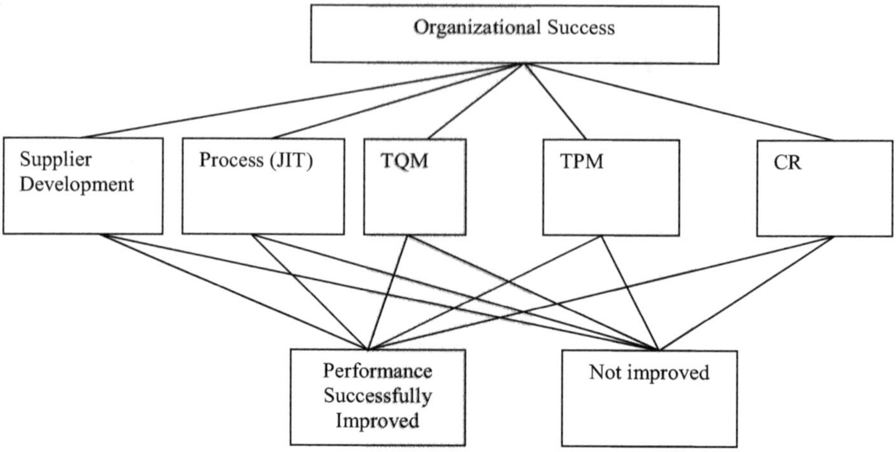

Fig. 6.1 Three-level hierarchy for improving the performance of manufacturing organization for strategic success

6.2.2 Step 6.2.2: Normalized Matrix of Different Sub-objectives

After a pair-wise comparison matrix is obtained, the next step is to divide each entry in column by the sum of entries in column to get the value of normalized matrix. The value of normalized matrix is shown in Table 6.2. The normalized value r_{ij} is calculated by the below mentioned formula:

$$r_{ij} = a_{ij} / \sum_{i=1}^{n} a_{ij}$$

Thus, the approximate priority weight ($W_1, W_2 .. W_j$) for each attribute is obtained as shown in Table 6.2.

Thus, the approximate priority weight (W_1, W_2, W_j) for each attribute is obtained as shown in Table 6.3.

$$W_j = 1/n \times \sum_{i=1}^{n} a_{ij}$$

6.2.3 Step 6.2.3: Do Consistency Check

The relative weights, which would also present the Eigen values of criteria, should be verified:

$$A \times W_i = \lambda_{max} \times W_i \ldots i = 1; 2; \ldots ; n$$

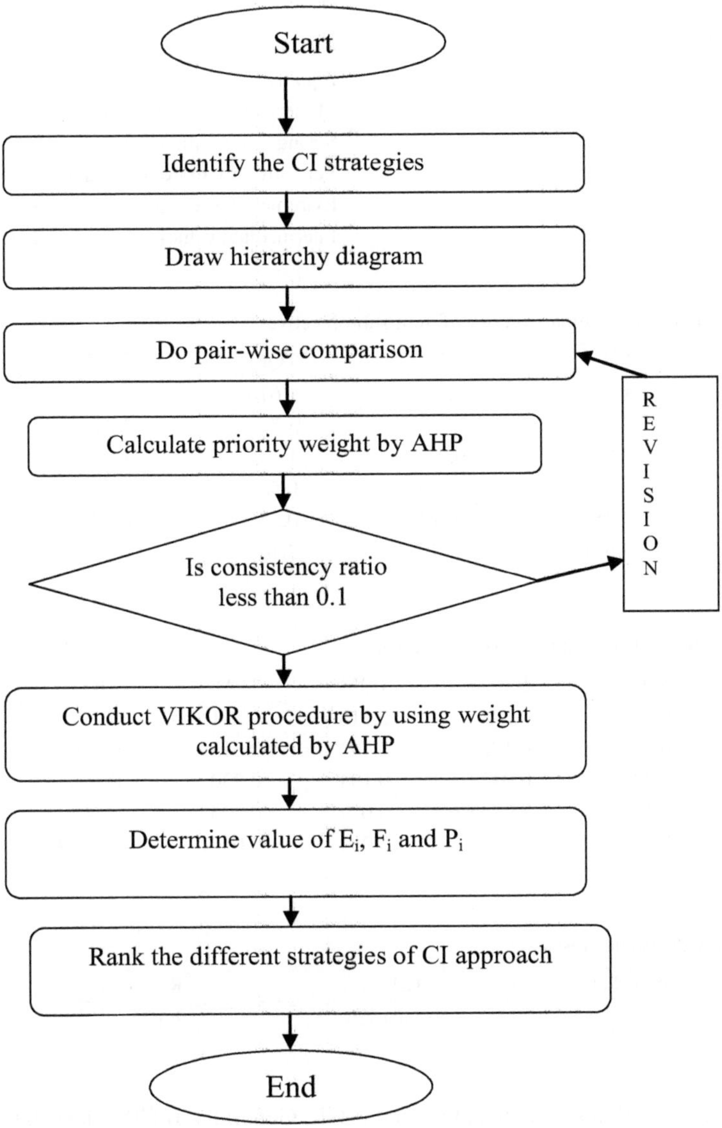

Fig. 6.2 Flowchart for methodology used

where A represents the pair-wise comparison decision matrix and $\lambda_{\max}$ gives the highest Eigen value. Then consistency index (CI), which measures the inconsistencies of pair-wise comparisons, is calculated as

$$\mathrm{CI} = \frac{(\lambda_{\max} - n)}{(n-1)}$$

Table 6.1 Degree of preferences

Definition	Intensity of importance
Equally important	1
Moderately more important	3
Strongly more important	5
Very strongly more important	7
Extremely more important	9
Intermediate values	2, 4, 6, 8

Table 6.2 Pair-wise comparison of different sub-objectives

	SD	JIT	TQM	TPM	CR
S.D	1	3.3	1.616	2	4
JIT	0.303	1	3.7	2.316	1.416
TQM	0.618	0.27	1	1.7	1.798
TPM	0.5	0.431	0.588	1	2.666
CR	0.25	0.706	0.556	0.375	1
Total	2.671	5.707	7.46	7.391	10.88

Table 6.3 Normalized matrix of sub-objectives

	SD	JIT	TQM	TPM	CR	Weight
SD	0.374	0.578	0.216	0.271	0.367	**0.3612**
JIT	0.113	0.175	0.49	0.313	0.130	**0.2442**
TQM	0.231	0.0473	0.134	0.230	0.165	**0.1672**
TPM	0.187	0.0755	0.077	0.135	0.245	**0.1439**
CR	0.0935	0.123	0.074	0.0507	0.0919	**0.0865**

Table 6.4 Consistency test

Maximum Eigen value	C.I	R.I	C.R
5.06107	0.01526	1.12	0.0136

The last ratio that has to be calculated is CR. Generally, if CR is less than 0.1, the judgments are consistent and acceptable. The formulation of CR is

$$CR = \frac{CI}{RI}$$

where random index (RI) denotes the average RI with the value obtained by different orders of the pair-wise comparison matrices. The values of consistency test are given in Table 6.4.

Table 6.5 Decision index table

		Improved performance	Not improved	Priority weight
SD	Improved	1	3.3	0.7674
	Not improved	0.3030	1	0.2326
JIT	Improved	1	3.64	0.7845
	Not improved	0.275	1	0.2155
TQM	Improved	1	1.7	0.6296
	Not improved	0.5882	1	0.3704
TPM	Improved	1	2.626	0.7246
	Not improved	0.38	1	0.2754
CR	Improved	1	1.366	0.5773
	Not improved	0.732	1	0.4227

6.2.4 *Step 6.2.4: Priority Weights for Alternatives with Respect to Attribute*

The chance of a successful CI strategic implementation increases if attribute present is strong. Priority weights for alternatives (improved and unimproved) are measured to show the preference of the alternative with respect to an attribute. Thus, if the presence of an attribute is strong in the organization, it is more likely to be a success, compared to the presence of an attribute which is weak. Table 6.5 summarizes the result of evaluating the possible outcome of the implementation with respect to each attribute.

The weight evaluation for each alternative is multiplying the matrix of evaluation rating by vector of attribute weight and summing over the entire attribute.

Decision index of improved performance: 0.7674 *0.3612 + 0.7845*0.2442 + 0.6296*0.1672 + 0.7246*0.1439 + 0.5773*0.0865 = 0.72811.

Thus, decision index of performance not improved = 1–0.73 = 0.27.

This signifies that the success rate of strategic implementation of CI approach is 73%.

6.3 Compromise Ranking Method

The VIKOR (the Serbian name is "Vlse Kriterijumska Optimizacija Kompromisno Resenje," which means multi-criteria optimization (MCO)) and compromise solution method was first established and later promoted by Zeleny (1982). It focuses on ranking and selecting the best alternative from a finite set of alternatives with conflicting criteria, and on proposing the compromise solution (one or more). The compromise solution is a feasible solution, which is the closest to the ideal solution, and a compromise means an agreement established by mutual concessions made between

Table 6.6 Values of E_i and F_i

$E_{SD} = 0.8472$	$F_{SD} = 0.3612$
$E_{JIT} = 0.6898$	$F_{JIT} = 0.2442$
$E_{TQM} = 0.24522$	$F_{TQM} = 0.16722$
$E_{TPM} = 0.3069$	$F_{TPM} = 0.1439$
$E_{CR} = 0.24932$	$F_{CR} = 0.08662$

the alternatives. The following multiple attribute merit for compromise ranking is developed from the Lp metric used in the compromise programming method:

$$L_{p,i} = \left\{\sum_{j=1}^{M} w_j \left[\left(m_{ij}\right)_{\max} - m_{ij}\right] / \left[\left(m_{ij}\right)_{\max} - \left(m_{ij}\right)_{\min}\right]^p\right\}^{1/p} \tag{6.1}$$

where M is the number of criteria and N is the number of alternatives. The m_{ij} values (for $i = 1, 2, \ldots, N; j = 1, 2, \ldots, M$) denote the values of criteria for different alternatives. In the VIKOR method, $L_{1,i}$ and $L_{\infty,j}$ are used to formulate the ranking measures.

6.3.1 Step 6.3.1: Determine the Values of $\mathbf{E}_i$ and $\mathbf{F}_i$

$$E_i = L_{1,i} = \sum_{j=1}^{M} w_j \left[\left(m_{ij}\right)_{\max} - m_{ij}\right] / \left[\left(m_{ij}\right)_{\max} - \left(m_{ij}\right)_{\min}\right] \tag{6.2}$$

$$F_{i=} L_{\infty,I=} \max{}^m \text{ of } \left\{w_j \left[\left(m_{ij}\right)_{\max} - m_{ij}\right] / \left[\left(m_{ij}\right)_{\max} - \left(m_{ij}\right)_{\min}\right]\right\} \tag{6.3}$$

The values of E_i and F_i are shown in Table 6.6. Acceptable Eq. (6.2) is only applicable to beneficial attributes (whose higher values are desirable). For non-beneficial attributes (whose lower values are preferable), the term $[(m_{ij})_{\max} - m_{ij}]$ in Eq. (6.2) is to be replaced by $[m_{ij} - (m_{ij})_{\min}]$. Hence, for non-beneficial attributes, Eq. (6.2) can be rewritten as

$$E_i = L_{1,i} = \sum_{j=1} w_j \left[m_{ij} - \left(m_{ij}\right)_{\min}\right] / \left[\left(m_{ij}\right)_{\max} - \left(m_{ij}\right)_{\min}\right] \tag{6.4}$$

6.3.2 Step 6.3.2: Calculation of $\mathbf{P}_i$ Values

$$P_i = v\left\{\left(E_i - E_{i-\min}\right) / \left(E_{i-\max} - E_{i-\min}\right) + (1 - v)\left(F_i - F_{i-\min}\right) / \left(F_{i\max} - F_{i-\min}\right)\right\}$$

where E_i-max and E_i-min are the maximum and minimum values of E_i, respectively, and F_i-max and F_i-min are the maximum and minimum values of F_i,

Table 6.7 Values of P_i and rank of attributes

Values of P_i	Rank
$P_{SD} = 1$	5
$P_{JIT = 0.6558}$	4
$P_{TQM} = 0.1468$	2
$P_{TPM} = 0.616$	3
$P_{CR} = 0.009$	1

respectively. v is introduced as weight of the strategy of "the majority of attributes" (or "the maximum group utility"). The value of v lies between 0 and 1. Normally, the value of v is taken as 0.5. The best alternative is the one having the minimum P_i value. The values of P_i and ranking of sub-objectives are shown in Table 6.7.

6.4 Methodology for Modeling

The learning issues drawn from literature, empirical research, and case studies conducted have been taken as options for a qualitative modeling. With a careful analysis, the overlapping and similar inferences have been scrutinized to develop a list of independent learning issues. Following this, the qualitative model has been developed by involving option field methodology (OFM), option profile methodology (OPM), analytic hierarchy process (AHP), and structural equation modeling (SEM).

6.4.1 Synthesis of Learning Issues

The learning issues of the empirical study and case studies have been synthesized and are given below:

The following three objectives, if met, make the organizations strategically successful for improving the performance of SMEs using CI strategies:

- Organization success (OS)
- Improved productivity (P)
- Improved quality (Q)

Status of different parameters of manufacturing organizations:

- Level of usage of CI strategies, viz. total quality management-related strategies
- Role of CI strategies for improving the performance of SMEs for strategic success
- Role of key enablers for improving the performance of SMEs

Different learning issues are as follows:

1. Significant increase in overall equipment effectiveness.
2. Poka-Yoke successfully implemented.
3. Improvement in market image of the SMEs.
4. Enhanced production rates.
5. Organizations use continuous improvement strategies, when its performance is proved.
6. Before using CI strategies, consequential small incremental changes in related process are considered.
7. Internal training and monitoring are to be imparted to the employees for improving overall operator efficiency.
8. Organizations have to redesign the organizational structure for implementing incremental changes.
9. Facing the problem in small incremental improvements.
10. Conducting customer awareness programs about strategic implementation of CI approach.
11. Pay full attention to every improvement activities.
12. Make a safety environment in the premises of the company.
13. Top management should chalk out long-term and short-term goals by using X-matrix.
14. Job rotation and job enrichment play a leading role in creating the interest of employee in job.
15. Improving the competitiveness of the organization.
16. Economic justification of incremental changes has to be determined.
17. Conducting customer awareness programs about changes in manufacturing system processes.
18. Best KAIZEN award system encourages employee towards improvement activities.
19. Reduction in setup time using SMED technique.
20. The organizations must take help of universities/technical colleges for suggesting the areas of improvement.
21. Formation of KAIZEN teams to expose every individual to different aspects such as planning, coordinating, and executing the ideas.
22. Rating of 5S activities is helpful in improving the system.
23. Determine the market needs regularly.
24. DMAIC approach is helpful in reducing rejections.
25. KAIZEN gallery approach is helpful to systemize the improvements.
26. Risk priority number (RPN) is significantly reduced by using FMEA strategy.
27. To create a vision of success while implementing CI strategies.
28. Availability of KAIZEN consultant helps in improving the processes.
29. Proper hierarchy is followed to successfully implement these strategies.
30. Proper understanding of current operations assists in strategic implementation of CI approach.
31. Increase in productivity.

32. Management support financially is one of the key factors for success of CI strategies.
33. Top management should chalk out long-term and short-term goals completely.
34. Make the employees aware about the role of CI activities.
35. Making small incremental changes to advance one's own capabilities.
36. Number of managers and technicians in the company are key factors responsible for implementation of CI strategies.
37. Active participation from engineers and technicians plays a positive role in CI strategic implementation.
38. Interactions between individuals within the organization are considered as one of the key factors for utilization of new technology.
39. Recruitment of new employees has been considered as important.
40. Higher education to employees is considered as important.
41. Proper work environment helps in utilizing the new technologies.
42. Advertising extensively with regard to quality and its features.

6.5 Different Methods for Modeling

A qualitative model using options field methodology (OFM), options profile methodology (OPM), analytic hierarchy process (AHP), and fuzzy set theory (FST) has been developed and applied to the situation to meet the objectives. The description of the model is given in Fig. 6.3.

6.5.1 Options Field Methodology (OFM)

1. To construct the polystructure: The completed options field is a polystructure. Its construction begins with the generation and classification of a set of options. This set may be generated by using modified idea writing in response to a

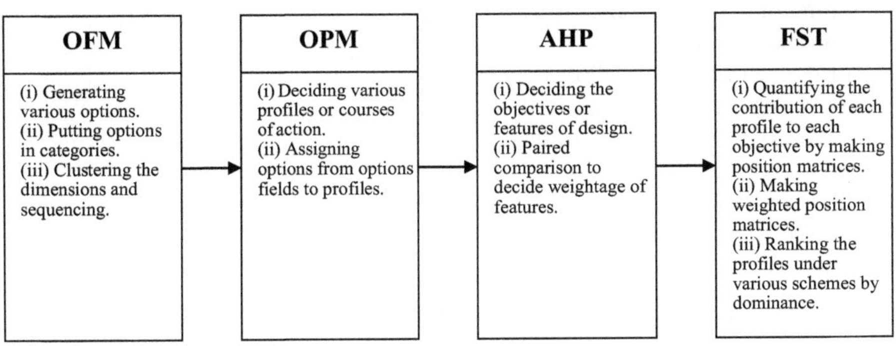

Fig. 6.3 Qualitative modeling

Fig. 6.4 Four-level structure quadrant

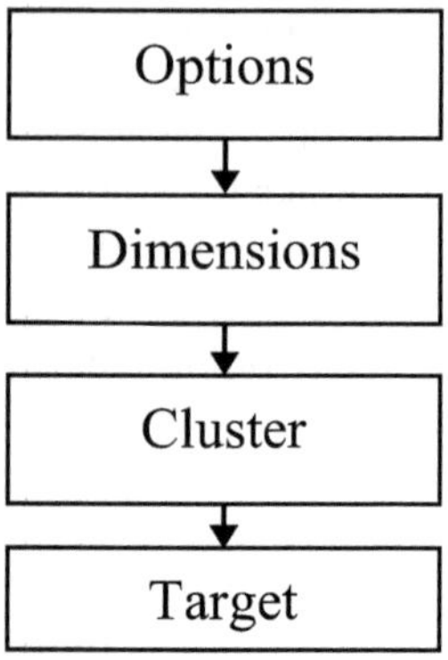

carefully formulated triggering question. This question defines the context and must, therefore, reflect substantial insight into the design situation. The question must neither be too broad nor too narrow. It must stimulate creative and productive responses that do not stray from the topic under consideration.

2. To initialize the structuring: Once a set is developed, the initial structuring begins. The initial structuring is for the purpose of placing options into categories. Options field is a triply structured quadrant, since it is a four-level structure, whose levels are named as options, dimension, cluster, and target reading from the top to bottom as shown in Fig. 6.4.

6.5.2 Formation of Categories

The modeling began with listing of options as a solution to the methodology of CI approach for strategic success. The list has been converted into a conceptual design. OFM/OPM was largely used as a basis for this purpose. Finally, the alternative options profiles developed were ranked using AHP. To name the categories: These options were then put into various categories. The categories are:

- Setting of strategic goals.
- Process planning.
- Formation of cross-functional teams.
- Human resource management.
- Sustained and consistent efforts.
- People (total involvement).
- Economic aspects.
- Cost-benefit analysis.
- Organizational priorities.

Learning issues are put into different categories as shown in Table 6.8.

Table 6.8 Formation of categories

S. no.	Category	Learning issues
1	Setting of strategic goals	20, 4, 31, 15, 24, 19, 26, 22, 27, 5
2	Process planning	30, 32, 33, 34, 35, 13, 23
3	Formation of cross-functional teams	36, 37, 38, 21, 25
4	Human resource management	7, 39, 40, 14, 18, 28
5	Sustained and consistent efforts	30, 12, 11, 9
6	People (total involvement)	41, 10, 17, 29
7	Economic actions	6, 16
8	Cost-benefit analysis	42, 1, 2
9	Organizational priorities	5, 8, 3

3. To identify the design dimensions: After the set of categories has been achieved, it is reasonable to believe that learning has occurred. All of these have been included and considered as the dimension of the design.
4. To discover the clusters of dependent dimensions: Once the group has settled on the dimensions of the target, a second structuring occurs. Now it is the set of dimensions that is structured. Two dimensions are defined to be independent if a choice of one or more options in one of the dimensions does not rule out any choices in the other dimension. If two dimensions are interdependent, the choice of options in one may be restricted by the choice of options in the other.
5. To establish a choice-making sequence for clusters: Now the third structuring begins. This structuring takes the clusters as elements to be structured. The structuring relationship involves the sequence in which choices of options should be made.
6. To sequence the dimensions within clusters: A fourth structuring is carried out now. In this, structuring is carried out separately for each cluster and initial decision-making sequence among dimensions in each cluster is defined. The dimensions were put into broader categories called clusters. Following are the clustering of the dimensions:

Improvement strategy

- Setting of strategic goals
- Economic actions
- Organizational priorities

Market responsiveness

- Process planning
- Cost-benefit analysis

Role of all concerned

- Formation of cross-functional teams
- People (total involvement)

Continuous upgradation

- Human resource management
- Sustained and consistent efforts

7. To display the completed options field: It is then appropriate to organize the options field by placing dimensions in the order determined with name of each dimension heading a list of options therein and with the cluster clearly identified.

6.5.3 *Options Profile Methodology (OPM)*

Options profile is the visual representation of an alternative consisting of a set of chosen options with at least one option coming from each dimension in the options field. Various profiles planned to improve the performance of SMEs are as follows:

- Process planning approach (PP)
- Advanced manufacturing technology (AMT)
- Human resource upgradation approach (HRU)
- Strategic approach (SA)

Having made the profiles, the next task is to list various objectives of the design or targets. Following this, contribution of each profile to each objective is determined by paired comparison. Analytical hierarchy process is employed for the purpose. A brief description of the AHP is given below:

Step 1: State the decision problem and overall objective/goal.
Step 2: Structure the hierarchy from the top through the intermediate to the lowest level, which usually contains a list of alternatives.
Step 3: Construct matrices of pair-wise comparisons for each of the lower level measurements. The decision-maker has the option of expressing intensity of preference on a nine-point scale values.
Step 4: Pair-wise comparisons among n elements in each level lead to an opportunity for evolving the structure into a reciprocal judgment matrix. Thus only $n(n-1)/2$ judgments are considered where n is the total number of elements.
Step 5: Calculate the consistency index (CI) is = $(\lambda_{max} - n)/(n-1)$ where n is the number of elements being compared and λ_{max} is the largest Eigen value of the judgment matrix. The smaller the value of CI, the smaller is the deviation from the consistency.
Step 7: Calculate the consistency ratio CR = CI/RI. Usually, a CR of 0.1 or less is considered as acceptable, and it reflects an informed judgment attributable to the knowledge of the analyst regarding the problem under study.

Various objectives of the organizations have been selected based on the analysis carried out earlier. The paired comparison method of analytic hierarchy process has been applied to find out the weightage of each objective. Each objective has been compared with each other independently by three respondents. These respondents

Table 6.9 AHP weight matrix: researcher

Criteria	OS	*P*	*Q*	Weights
OS	1	3	2	0.512
P	1/3	1	4	0.328
Q	1/2	1/4	1	0.158

Table 6.10 AHP weight matrix: manager 1

Criteria	OS	*P*	*Q*	Weights
OS	1	2.5	2.83	0.55
P	0.4	1	2.67	0.31
Q	0.353	0.374	1	0.15

Table 6.11 AHP weight matrix: manager 2

Criteria	OS	*P*	*Q*	Weights
OS	1	2.83	2.5	0.549
P	0.353	1	2.67	0.291
Q	0.4	0.375	1	0.159

Table 6.12 Weights of various objectives

Objective respondent	OS	*P*	*Q*
Researcher	0.512	0.328	0.158
Manager 1	0.55	0.31	0.15
Manager 2	0.549	0.291	0.159

were chosen to reduce the biases. Out of three respondents chosen, one was quality manager from M/S Aarti Steels Limited, Ludhiana; second one was senior manager from M/S G S Auto International Limited, Ludhiana; and third one has been the researcher himself. The respondents compared the objectives on a qualitative scale as depicted in Tables 6.9, 6.10, 6.11, and 6.12.

6.5.4 *Fuzzy Set Theory*

Fuzzy sets and subsets arc those ill-specified and non-distinct collection of objects with unsharp boundaries in which transition from membership to nonmembership is gradual rather than abrupt. A fuzzy set is characterized by a membership function, defined as a real number in the interval (0, 1) (see Table 6.13). The fuzzy set methodology for multi-criteria decision-making method (MCDM) is used to analyze various options.

After determining the weights of the objectives, the next step has been to make position matrices. In order to represent the views of each of the interest group, a position matrix is prepared from the responses of all the experts in the group by giving numerical

Table 6.13 Conversion of linguistic terms into fuzzy scores

Exceptionally low	0.045
Extremely low	0.135
Very low	0.255
Low	0.335
Below average	0.410
Average	0.500
Above average	0.590
High	0.665
Very high	0.745
Extremely high	0.865
Exceptionally high	0.955

Table 6.14 Different position matrices

Profile objectives	Process planning approach (SP)	Continuous improvement approach (CI)	HR upgradation approach (HRU)	Marketing approach (MA)	Weight
Position matrix: researcher					
OS	0.865	0.590	0.665	0.745	0.164
P	0.745	0.500	0.410	0.335	0.539
Q	0.745	0.590	0.335	0.665	0.297
Position matrix: manager 1					
OS	0.745	0.500	0.590	0.865	0.196
P	0.665	0.500	0.335	0.335	0.493
Q	0.865	0.590	0.410	0.590	0.311
Position matrix: manager 2					
OS	0.865	0.500	0.590	0.745	0.168
P	0.590	0.590	0.335	0.410	0.483
Q	0.665	0.590	0.500	0.665	0.349

values to the qualitative assessment. Average value of each element representing the group response is worked out by multiplying membership function value of each alternative as given by the respondents with assigned weight, i.e., the Eigen vector weight as determined by AHP. This way some of the bias in the matrix can be eliminated. The weighted matrices for each of the interest group are thus prepared. There are three ways to aggregate the weighted matrix, viz. optimistic, average, and pessimistic aggregation. The highest value among various group responses represents the optimistic value, the lowest value represents the pessimistic value, and the average of all the values represents the mean value. Again, this exercise has been done by the three respondents. The position matrices along with the weights determined earlier are given in Table 6.14.

From the position matrices, weighted position matrices have been determined. This has also been done individually for the matrix from each respondent. The weight of the objective as determined earlier has been multiplied by the value of each position of the position matrix and weighted values have been obtained. Table 6.14 shows the weighted position matrices.

Table 6.15 Optimistic weighted position matrix

Profile objectives	Process planning approach (SP)	Advanced manufacturing technology (CI)	HR upgradation approach (HRU)	Marketing approach (MA)
OS	0.146	0.098	0.116	0.170
P	0.402	0.285	0.221	0.198
Q	0.269	0.206	0.175	0.232

Table 6.16 Pessimistic weighted position matrix

Profile objectives	Process planning approach (SP)	Advanced manufacturing technology (AMTs)	HR upgradation approach (HRU)	Marketing approach (MA)
OS	0.142	0.084	0.099	0.122
P	0.285	0.247	0.162	0.165
Q	0.221	0.175	0.099	0.158

Table 6.17 Average weighted position matrix

Profile objectives	Process planning approach (SP)	Advanced manufacturing technology	HR upgradation approach (HRU)	Marketing approach (MA)
OS	0.144	0.093	0.108	0.139
P	0.338	0.267	0.183	0.181
Q	0.241	0.188	0.134	0.191

The best course of action for a decision-maker in any situation is to use a Hadley's criteria of cautious optimism. If "*A*" is the dominance weight of the option as determined from optimistic matrix and "*B*" that of the pessimistic dominance matrix, weight of the option according to Hadley's criterion is determined by the relationship:

$$W = \alpha \times A + (1 - \alpha) \times B$$

Appendix XII shows the Hadley's weighted position matrices for different levels. From these weighted position matrices, optimistic, pessimistic, and average weighted position matrices have been made using FST. For optimistic matrix, the highest of each position has been selected, for pessimistic the lowest values have been selected, and for average the average values have been selected. Tables 6.15, 6.16, and 6.17 show these values.

Following this, dominance matrices have been prepared. In these matrices, the dominance of each course of action over the others has been tabulated. Dominance matrix is prepared to display the dominance structure between all possible pairs of options The element d_{ij} is the number of features for which membership value of option *j* dominates or is greater than option *i*. A dash is entered for the diagonal d_{ij} element. If the *K*th column is summed, the total number of dominances of option *K* over all options is obtained. Similarly, if the *K*th row is summed, the number of times the *K*th option is

Table 6.18 Dominance matrix—optimistic

	PP	AMT	HRU	MA	Row sum	Rank
PP	–	0	0	1	1	I
AMT	3	–	1	2	6	III
HRU	3	2	–	2	7	IV
MA	2	1	1	–	4	II
Column sum	8	3	2	5		
Rank	I	III	IV	II		

Table 6.19 Dominance matrix—pessimistic

	PP	AMT	HRU	MA	Row sum	Rank
PP	–	0	0	0	0	I
AMT	3	–	1	1	5	II
HRU	3	2	–	3	8	III
MA	3	2	0	–	5	II
Column sum	9	4	1	4		
Rank	I	II	III	II		

Table 6.20 Dominance matrix—average

	PP	AMT	HRU	MA	Row sum	Rank
PP	–	0	0	0	0	I
AMT	3	–	1	2	6	III
HRU	3	2	–	2	7	IV
MA	3	1	1	–	5	II
Column sum	9	3	2	4		
Rank	I	III	IV	II		

being dominated by all other options is determined. More favorable outcomes have higher column sums and lower row sums. The cell value denotes that a course of action dominates other courses of action in how many criteria and it is dominated by another course of action in how many criteria. Thus row sum depicts the number by which a criterion is dominated and the column sum depicts the number by which the profile dominates all other profiles. The matrices are presented in Tables 6.18, 6.19, 6.20, and 6.21 shows the Hadley's dominance matrices for different levels.

6.5.4.1 Discussion of the Findings

The dominance matrix has shown that process planning has emerged as the preferred strategy for all the profiles. Marketing, advanced manufacturing technology, and human resource upgradation strategies have been placed at second, third, and fourth positions, respectively, in 100%, 80%, and 50% and average level of optimism. However, in case of 100% pessimistic level, strategic planning is placed at rank one. Advanced manufacturing technology and marketing one has been come at rank two.

Table 6.21 Dominance matrix of different optimistic levels

Rank profile	100% optimistic	80% optimistic	50% optimistic	20% optimistic	100% pessimistic	Average
SP	I	I	I	I	I	I
AMT	III	III	III	II	II	III
HRU	IV	IV	IV	IV	III	IV
MA	II	II	II	III	II	II

The rank goes to human resource upgradation. Almost similar results have been obtained at 20% optimism level.

From the above, it can be inferred that the foremost attention has to be given to the strategic planning. The timing development to prepare the strategic planning would help organizations to improve the performance. The second strategy is marketing approach. This strategy should go parallel to the manufacturing of different products. Third strategy is the continuous improvement in the whole organization. CI projects will the help the organizations in improving the performance of organizations at each level. No company can ignore the role of human beings in any type of activities. It is expected that the people must be acquainted with all the roles to enhance the performance of firm.

6.6 Structural Equation Modeling

Structural equation modeling is a family of statistical models that seek to explain the relationship among multiple variables. To do so, it examines the structure of interrelationships expressed in a series of equations, similar to a series of multiple regression equations. These equations depict all of relationship among construct involved in the analysis. Constructs are unobservable or latent factors that are represented by multiple variables. In Fig. 6.5, CI represents continuous improvement system and structure. FP represents firm performance, SP represents supplier, and customer relationship and OC represent organization culture. H1, H2 and H3, and H4 and H5 indicate the hypotheses.

CI, OC FP, and SR signify latent variables. Positive correlation coefficients between the various latent variables including CI, OC, FP, and SR suggest the positive hypotheses for the research as shown in Table 6.22.

The hypotheses have been framed for the present research in concern with structural equation modeling as shown below:

- H1: Continuous improvement system and structure are positively associated with firm performance.
- H2: Organizational culture is positively associated with firm performance.
- H3: Continuous improvement system and structure are positively associated with organizational culture.
- H4: Continuous improvement system and structure are positively associated with supplier and customer relations.

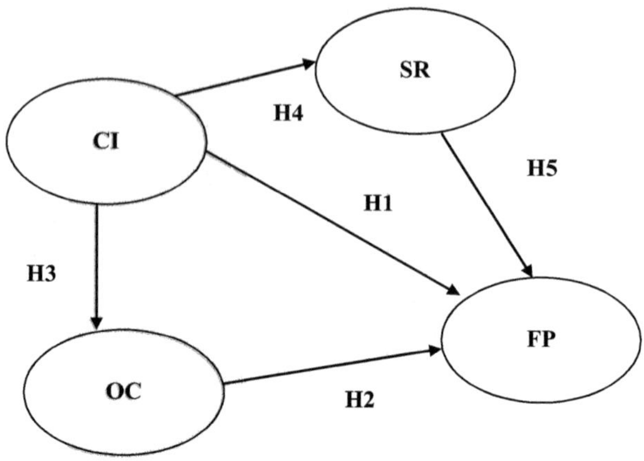

Fig. 6.5 Research framework and hypotheses

Table 6.22 Correlation between various variables

	CI	SR	OC	FP
CI	1	0.01605	0.147516	0.093104
SR		1	0.091741	0.144956
OC			1	0.12254
FP				1

Table 6.23 shows the various indices of the proposed model

Indices	Calculated values	Standard values
Chi-square/degree of freedom	1.4843	Value should be less than 3
GFI (goodness of fit index)	0.89	$0 \leq GFI \leq 1$
CFI (comparative fit index)	0.931	$0 \leq CFI \leq 1$
TLI (Tucker-Lewis index)	0.895	$0 \leq TLI \leq 1$
AGFI (adjusted goodness of fit)	0.875	$0 \leq AGFI \leq 1$
Probability level	0.0001	<0.05

- H5: Supplier and customer relations are positively associated with firm performance.

Table 6.23 indicates the high model conformity prepared for improving the performance of SMEs. The variables assigned to each of the dimensions have been subjected to factor analysis to ensure that they are reliable indicators of those constructs. A cutoff loading of 0.5 has been used to screen out variables that are weak indicators of the constructs. The reliability coefficient of different variables is greater than 0.7 which is adequate for every empirical research as shown in Table 6.24.

Table 6.24 Factor loadings and reliability coefficient of different variables

S. no.	Variable	Factor loading	Cronbach's alpha
Continuous improvement system and structure			
1	Employees search for information, new ideas, and technologies as a part of CI practices (c1)	0.689	
2	Employees facilitate and encourage informal relationship as part of continuous improvement practices (c2)	0.780	
3	Employees take reasonable risks by continuously experimenting with the new ways of doing work (c3)	0.561	0.789
4	Employees use failures as opportunity to learn as part of continuous improvement (c4)	0.619	
5	Employees let organizational objectives guide the evaluation of new ideas and information as part of continuous improvement practices (c5)	0.613	
6	Employees actively monitor progress by using action plans to ensure the goals of continuous improvement program (c6)	0.544	
Customer and supplier relationship			
1	The firm's reputation is important to the firm's competitive advantage (s1)	0.619	
2	The product supplied is important to the firm's competitive advantage (s2)	0.713	0.815
3	The customer satisfaction is important in the firm's competitive advantage (s3)	0.655	
Organizational culture			
1	The organizational culture encourages employees to hold knowledge closely (o1)	0.531	
2	The organizational culture reinforces behavior that upholds traditions (o2)	0.680	0.791
3	The culture encourages managers to closely monitor work time (o3)	0.691	
4	The culture focuses on short-term performance (o4)	0.731	
5	The culture encourages employee to interact with other people (o5)	0.612	
	Firm performance		
1	As a result of continuous improvement in product, faster speed to market occurs based on perception of actual performance (f1)	0.719	
2	As a result of continuous improvement in product, the higher product configurations occur based on perception of actual performance (f2)	0.753	0.735
3	As a result of continuous improvement in product, higher success of new products launched occurs based on perception of actual performance (f3)	0.745	
4	Improved product innovation of new parts and processes occurs based on perception of actual performance (f4)	0.642	

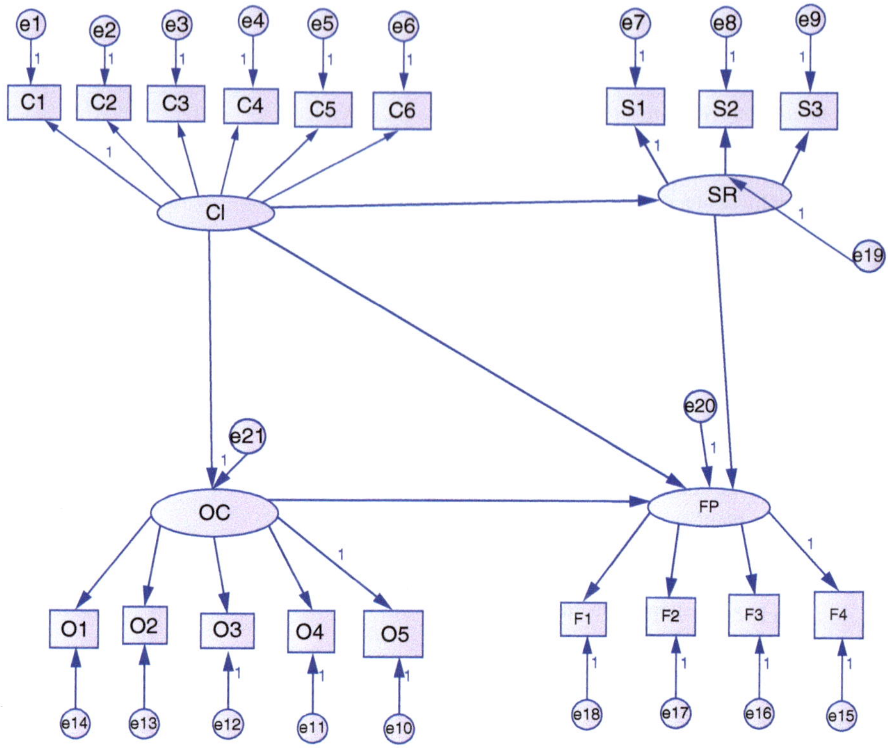

Fig. 6.6 Path diagram depicting relationship between endogenous and exogenous variables (mixed model)

6.6.1 Measurement Model for Improving the Performance of SMEs

To assess the performance of SMEs, an individual model and mixed model has been prepared to ascertain the relationship between latent variables with each other and also measured variables assigned to each latent variable. The error variables depicting measurement error are also included in relation to measurement variables. The variance of all the error variables has been fixed in software AMOS 22. The standard path coefficient (SPC) of one individual measurement is fixed and the value of other path coefficients is measured relative to fixed measurement variable having SPC unity. Figure 6.6 indicates the individual measurement model showing individual latent variables and measurement variable model and Fig. 6.6 shows mixed model depicting endogenous (variables that are influenced by other variables in a model) and exogenous (variables that are not influenced by other variables in a model) variables. In the present study continuous system and structure are the exogenous variables and other three variables including organizational culture, customer and supplier relationship, and supplier and customer relations are

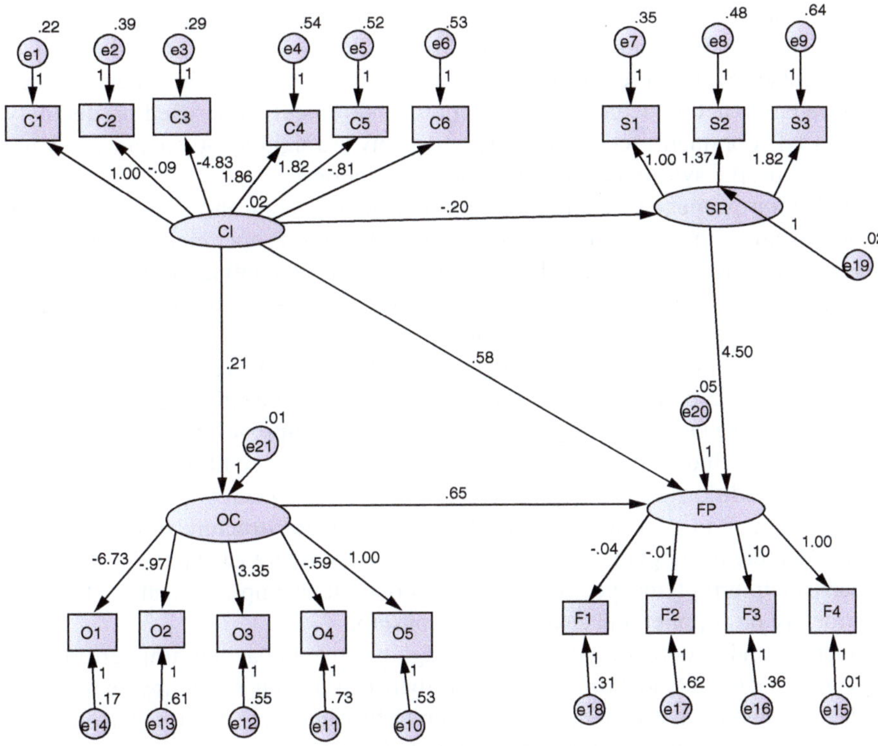

Fig. 6.7 Measurement model of improving the performance of SME

endogenous variables. Figure 6.7 indicates mixed model depicting relationship between various variables including both endogenous variables and exogenous variables and depicts the relationship of measurement variables and latent variables.

6.6.2 Validation of Hypotheses Framed

This study has provided empirical justification for the proposed research model which investigates the relations between continuous improvement system and structure, organization culture, and customer and supplier relations on firm performance among the SMEs of Northern India. The results of investigation demonstrate that proposed structural model closely fits the sample data and hypotheses are tested in this research. The studies assess the impact of continuous improvement system and structure, customer and supplier relation, and organizational culture on firm performance. Random measurement errors and error uniqueness, terms used to describe an error variance arising from some characteristics, are considered to be specific to a particular indicator variable. Residual terms represent error in the prediction of endogenous factors from exogenous factors.

- This result suggests that for improvement in continuous system and structure, firm performance is increased by 0.58 times. Following the proper hierarchy of organization structure is positively associated with the firm performance. Different benefits including organization achievements, improved productivity, reduced poor quality, cost, and increased safety are significantly achieved by following proper system and structure of organization.
- With improvement in continuous improvement system and structure there is improvement in organizational culture by 0.21 times. Proper hierarchy and organization structure significantly affect the culture of continuous improvement. The culture of continuous improvement is easily achieved by proper utilization of system and structure of the organization.
- With improvement in organizational culture, there is improvement in firm performance by 0.65 times. Improper culture within the organization adversely affects the performance of SMEs. Different parameters improve the performance of SMEs in the Indian context. The organizational culture reinforces behavior that upholds traditions of incremental changes in the organization.
- Improvement in continuous improvement system and structure shows considerable decline in supplier and customer relationship by a factor of 0.20. Different barriers interrupt the positive effect of continuous improvement system and structure on supplier and customer relationship.
- Supplier and customer relationship has high positive impact on firm performance with the factor of 4.50. Proper relationship between customer and supplier improves the coordination between these two which further improves the firm performance.
- Employees use failures as an opportunity to learn as part of continuous improvement that improves the firm performance by a factor of 1.056. Failures coming in the path of incremental improvements allow employee to learn a lesson which leads to improvement in firm performance.
- With the increase in employee search for information, new ideas and technologies as a part of CI practices aim at improving firm performance by .58 times. Employees search for information and ideas of system improvement lead to improvement in firm performance.
- Employees use failures as an opportunity to learn as a part of continuous improvement which shows positive influence on firm performance by a factor of 1.056. Failure in continuous improvement activities allows employees to learn how to improve the overall operator efficiency and improve the performance of manufacturing system processes.
- The customer satisfaction is important in improving firm's competitive advantage that plays a significant role in improving the performance of firm by a factor of 8.19. Customer satisfaction regarding delivery, cost, and quality improves the competitive image of the organization which further leads to improvement in performance of organization.
- The culture of continuous improvement encourages managers to closely monitor work time which increases the performance of manufacturing organization by a factor of 2.176. Continuous improvement activities performed by managers play

an important role in monitoring work time, which further leads to improvement in firm performance.

- As a result of continuous improvement in product, higher success of new products launched occurs based on perception of actual performance which improves the firm performance by a factor of .10. This signifies that launching of new product improves the performance of manufacturing operations and hence improves the reliability of continuous improvement activities.
- The product supplied is important for the firm's competitive advantage which plays an important role in improving the performance of current manufacturing system processes by a factor of 6.165. This indicates that timely supply of product to the customer improves the performance of the organization.
- Employees facilitate and encourage informal relationship as part of continuous improvement practices which has a negative effect on firm performance by a factor of 0.0522. This indicates that informal relationship between various employees of the organization is the hindrance in improving the performance of organization.
- The culture focused on short-term performance negatively affects the firm performance by a factor of 0.3835. SMEs are focused on long-term performance to improve the performance of manufacturing system processes.

Chapter 7
Conclusions, Limitations, and Scope for Future Work

The research has led to some significant revelation about various facets of improving performance of SMEs in Northern India. The following conclusions are drawn from this study:

7.1 Conclusions

7.1.1 Conclusions from the Survey

(a) Continuous improvement strategies including 5S, total cost management, preventive maintenance, quality function deployment, suggestion system, finance, marketing, and policy deployment can be used to perform small incremental changes in their manufacturing system processes.
(b) These organizations are using total quality management strategies at the highest maturity level.
(c) CI strategies including value stream analysis, redesign, suggestion system, failure mode effect analysis, process flow mapping, recognition, minor stoppage elimination, and customer quality, cost, and delivery analysis (QCD) through holistic CI implementation program can significantly contribute to improvements in the organization.
(d) Strategic implementation of CI approach significantly contributes towards improving the quality of the product manufactured by the SMEs.
(e) Strategic implementation of CI approach has positive as well as negative impact on different performance parameters.
(f) CI approach through manufacturing performance success is helpful in realizing significant manufacturing benefits in Indian SMEs.

J. Singh, H. Singh, *Strategic Implementation of Continuous Improvement Approach*, SpringerBriefs in Operations Management,
https://doi.org/10.1007/978-3-319-93121-0_7

(g) Continuous improvement process declines in maturity phase as compared to introductory and stabilization phases.

(h) KAIZEN approach is highly helpful towards cost optimization by focusing on addressing overhead expense-related issues by small incremental improvements, stabilizing quality systems by reducing operator mistakes, facilitating improvements in the production system by improving equipment effectiveness, providing customer satisfaction through dependable and timely delivery of the product, maintaining health and safety standards as recommended in the safety policies, and achieving goals of the organization by improving the competitive image among the competitors.

(i) Significant improvement in productivity can be realized by exhibiting improved cost control over manufacturing facilities, stabilizing production systems, and affecting improvements in the cost control systems over the entire equipment life cycle.

(j) CI activities through strategic implementation can contribute towards quality enhancements by eliminating deteriorations in production systems but also has a negative impact on the safety-related issues of employees.

(k) CI approach helps organizations to achieve organization goals by reducing costs associated with the manufacturing system processes.

(l) CI enablers contribute significantly both in positive and negative sense towards the realization of strategic manufacturing performance parameters for competing in this highly competitive world.

(m) Customer relationship plays a significant role in improving the performance of the organization and success rate of strategic implementation of CI approach is about 73%.

(n) It is proved that all the five parameters are important for the improved performance of SMEs, although in varying degrees that is evidenced by AHP and VIKOR methodology. Customer relationship is rated the most important strategy among all.

(o) Inadequate knowledge or understanding of KAIZEN is the most significant barrier in implementing CI activities.

(p) SMEs must exploit all available CI strategies to supplement manufacturing resource planning for greater quality of the product.

7.1.2 Conclusions from the Case Studies

(a) CI strategic implementation has a huge potential in effecting significant improvement in overall equipment effectiveness that can also effectively contribute towards realization of significant productivity enhancements.

(b) It prepares the organizations to meet the challenges of competitive manufacturing by effective implementation of a well-conceived plan with the help of autonomous maintenance.

(c) The case studies have revealed that the Indian entrepreneurs have reaped significant manufacturing improvement through improved process reliability like OEE, affecting cost reductions, thereby strengthening sustainability efforts of the organization to meet competitiveness.
(d) The study reported in this work has revealed that there is an emerging need for CI strategic implementation in the Indian farm parts industry and the need to develop a sustainable action plan to foster CI implementation practices.
(e) The study has outlined the significance of continuous improvement strategies as a vital factor for achieving goals of the organization. It therefore becomes pertinent for the successful organizations to integrate improvement initiatives by reduction of setup time using SMED approach.
(f) Management should encourage the employees regularly to improve the performance of SMEs using CI strategies.
(g) CI implementation programs like preventive maintenance, policy deployment, and small-group activities can facilitate the manufacturing organization's quest for achieving enhanced manufacturing performance.
(h) Imparting on-job and off-job training to the employees significantly increases overall operator efficiency.
(i) The organizations must develop behavioral skills of facilitators to foster and manage culture change while preparing them to focus teams on a process for achieving continuous improvement.
(j) CI programs are strategically focused on inculcating team work and deployment of self-directed cross-functional teams in the organization, since teams play a crucial role in improving equipment effectiveness through increased mean time between failures.

7.1.3 Conclusions from Implementation Plan

(a) SEM implies that customer and supplier relation plays a highly significant role in improving the performance of firm and hence improves the manufacturing system processes showing high positive correlation.
(b) The organization success has been achieved after successful implementation of CI approach.
(c) Poor continuous improvement system and structure show considerable decline in supplier and customer relationship.
(d) ISM implies that barriers including lack of management commitment, lack of continuous training and education to employees, lack of employee trust in senior management, and inadequate knowledge of KAIZEN have high driving power and low dependency power which significantly affect the performance of SMEs.

7.2 Limitations of the Study

The study has the following limitations:

(a) Generalization of CI strategies is limited, but the study does contribute to the literature on further development of sophisticated CI methodology like Lean Six Sigma in organizations.
(b) Moreover the study is limited only to Northern Indian SMEs.
(c) Convenience sampling technique is used.
(d) The study is limited in inclusion of additional issues related to theoretical questions in achieving performance excellence because of some literature constraints and requirements.
(e) Limited number of strategies has been implemented to ascertain the important benefits occurred after strategic implementation of CI approach.
(f) There was only one respondent per company, so there is a possibility of method variance.

7.3 Scope for Future Work

(a) This survey can be extended to service organizations as well.
(b) The survey can be extended by determining the level of difficulties in implementing these strategies and the reasons for these difficulties.
(c) The analysis can be expanded on the basis of which it could be recommended that the manufacturing organizations should implement the most important and least difficult elements of KAIZEN.

Appendix: KAIZEN Questionnaire

General Organizational Information

Organization's name		
Organization's address		
Respondent's name and designation		
Respondent's contact no./e-mail address		
Primary product(s) of the organization		
Type of company	Small scale	Medium scale

Since when you have started your manufacturing organization	1–4 years	5–8 years	9–12 years	More than 12 years
Since when you are applying CI approach	<2 years	2–4 years	4–6 years	>6 years
What is the present status of implementing CI strategies	Early stages	Midway stages	Majority completed	Successfully completed
Which type of KAIZEN is more effective	Process type		Flow type	

Part A: Please indicate the level of usage (maturity level) and level of importance of continuous improvement strategies in your organization

	Level of usage				Level of importance			
CI strategies	Mature (1)	Developing (2)	Immature (3)	Not applicable (4)	Not at all important (1)	To some extent (2)	To a moderate extent (3)	To a large extent (4)
Supplier development								
Total cost management	1	2	3	4	1	2	3	4
Value analysis	1	2	3	4	1	2	3	4

J. Singh, H. Singh, *Strategic Implementation of Continuous Improvement Approach*, SpringerBriefs in Operations Management,
https://doi.org/10.1007/978-3-319-93121-0

	Level of usage				Level of importance			
CI strategies	Mature (1)	Developing (2)	Immature (3)	Not applicable (4)	Not at all important (1)	To some extent (2)	To a moderate extent (3)	To a large extent (4)
Value stream analysis	1	2	3	4	1	2	3	4
Process (just in time)								
Cycle time reduction	1	2	3	4	1	2	3	4
Process flow analysis	1	2	3	4	1	2	3	4
Material flow analysis	1	2	3	4	1	2	3	4
Theory of constraints	1	2	3	4	1	2	3	4
Redesign	1	2	3	4	1	2	3	4
Workflow analysis	1	2	3	4	1	2	3	4
Total quality management								
5S	1	2	3	4	1	2	3	4
Error proofing analysis	1	2	3	4	1	2	3	4
Failure mode effect analysis	1	2	3	4	1	2	3	4
Six Sigma	1	2	3	4	1	2	3	4
Standardization	1	2	3	4	1	2	3	4
Total productive maintenance								
Preventive maintenance	1	2	3	4	1	2	3	4
Autonomous maintenance	1	2	3	4	1	2	3	4
Equipment restoration	1	2	3	4	1	2	3	4
Change over reduction: SMED	1	2	3	4	1	2	3	4
Minor stoppage elimination	1	2	3	4	1	2	3	4
Customer relationship								
Quality function deployment	1	2	3	4	1	2	3	4
Customer quality, cost, and delivery analysis (QCD)	1	2	3	4	1	2	3	4
If any other, please specify								

Part B: Please indicate the level of usage (maturity level) and level of importance of CI support methods in your organization

CI support methods	Mature (1)	Developing (2)	Immature (3)	Not applicable (4)	Not at all important (1)	To some extent (2)	To a moderate extent (3)	To a large extent (4)
	People (total involvement)							
Internal training and monitoring	1	2	3	4	1	2	3	4
Manager development	1	2	3	4	1	2	3	4
Suggestion system	1	2	3	4	1	2	3	4
Team-based improvement	1	2	3	4	1	2	3	4
Principles of KAIZEN	1	2	3	4	1	2	3	4
	System (support core work)							
Finance	1	2	3	4	1	2	3	4
Marketing	1	2	3	4	1	2	3	4
Process flow mapping	1	2	3	4	1	2	3	4
Support and administrative KAIZEN	1	2	3	4	1	2	3	4
	Leadership							
Vision alignment and direction	1	2	3	4	1	2	3	4
Policy deployment	1	2	3	4	1	2	3	4
Recognition	1	2	3	4	1	2	3	4
	Promoting culture							
Use of ISO 9000/TS Standards	1	2	3	4	1	2	3	4
Promoting through notice boards	1	2	3	4	1	2	3	4
Promoting through internal media (magazines, newsletters)	1	2	3	4	1	2	3	4
Regular shop floor visit by management	1	2	3	4	1	2	3	4
Face-to-face communication	1	2	3	4	1	2	3	4
If any other, please specify								

Part C: Please tick mark (√) the appropriate choice to indicate the important barriers in implementing CI activities

S. No.	Barriers	Not at all important (1)	Neutral (2)	Important (3)	Most important (4)
1	Poor planning	1	2	3	4
2	Lack of management commitment	1	2	3	4
3	Resistance of the workforce	1	2	3	4
4	Lack of continuous training and education	1	2	3	4
5	Teamwork complacency	1	2	3	4
6	Use of an off-the-shelf program	1	2	3	4
7	Failure to change organizational philosophy	1	2	3	4
8	Lack of resources	1	2	3	4
9	Inability to change organizational culture	1	2	3	4
10.	Inability to build a learning organization that provides for continuous improvement	1	2	3	4
11	Incompatible organizational structure and isolated individuals and departments	1	2	3	4
12	Inappropriate reward system	1	2	3	4
13	Ineffective measurement techniques	1	2	3	4
14	Paying inadequate attention to customers	1	2	3	4
15	Inadequate use of empowerment and teamwork	1	2	3	4
16	Poor interdepartmental relations	1	2	3	4
17	Lack of employee trust in senior management	1	2	3	4
18	Politics and turf issues	1	2	3	4
19	Lack of strong motivation	1	2	3	4
20	Drive for short-term financial results	1	2	3	4
21	Lack of leadership	1	2	3	4
22	Lack of time to devote to quality initiative	1	2	3	4
23	Incompatible organizational structure and isolated individuals and departments	1	2	3	4
24	Use of a prepackaged program	1	2	3	4
25	Inadequate knowledge or understanding of KAIZEN	1	2	3	4
26	If any more, please specify				

Parts D: Please indicate the role of CI values in enhancing the performance of continuous improvement strategies

Values	Not at all important 1	To some extent 2	To a moderate extent 3	To a great extent 4
		CI driving values		
Respect	1	2	3	4
Responsibility	1	2	3	4
Empathy	1	2	3	4
		CI enabling values		
Humbleness	1	2	3	4
Trust	1	2	3	4
Openness	1	2	3	4
Cooperation	1	2	3	4
		CI infusion values		
Leadership				
Structure support	1	2	3	4
Control system	1	2	3	4
Explicit behavior change	1	2	3	4
		Customer relationship		
Efficiency	1	2	3	4
Quality	1	2	3	4
Competence	1	2	3	4
Flexibility	1	2	3	4
Innovation	1	2	3	4
Speed	1	2	3	4
Capacity	1	2	3	4
Uniqueness	1	2	3	4
Reliability	1	2	3	4

Part E: To what extent do you agree with the following statements concerning the variable for enhancing the performance of continuous improvement strategies?

S. No.	Variables	Disagree (1)	Neutral (2)	Agree (3)	Strongly agree (4)
	Continuous improvement system and structure				
1	Employees search for information, new ideas, and technologies as a part of CI practices	1	2	3	4
2	Employees facilitate and encourage informal relationship as part of continuous improvement practices	1	2	3	4

S. No.	Variables	Disagree (1)	Neutral (2)	Agree (3)	Strongly agree (4)
3	Employees take reasonable risks by continuously experimenting with the new ways of doing work	1	2	3	4
4	Employees use failures as an opportunity to learn as part of continuous improvement	1	2	3	4
5	Employees let organizational objectives guide the evaluation of new ideas and information as part of continuous improvement practices	1	2	3	4
6.	Employees actively monitor progress by using action plans to ensure the goals of continuous improvement program	1	2	3	4
	Customer and supplier relationship				
1	The firm's reputation is important to the firm's competitive advantage	1	2	3	4
2	The product supplied is important to the firm's competitive advantage	1	2	3	4
3	The customer satisfaction is important in the firm's competitive advantage	1	2	3	4
	Organizational culture				
1	The organizational culture encourages employees to hold knowledge closely	1	2	3	4
2	The organizational culture reinforces behavior that upholds traditions	1	2	3	4
3	The culture encourages managers to closely monitor work time	1	2	3	4
4	The culture focuses on short-term performance	1	2	3	4
5	The culture encourages employee to interact with other people	1	2	3	4
	Firm performance				
1	As a result of continuous improvement in product, faster speed to market occurs based on perception of actual performance	1	2	3	4
2	As a result of continuous improvement in product, the higher product configurations occur based on perception of actual performance	1	2	3	4
3	As a result of continuous improvement in product, higher success of new products launched occurs based on perception of actual performance	1	2	3	4
4	Improved product innovation of new parts and processes occurs based on perception of actual performance	1	2	3	4

Part F: To what extent your organization has achieved the benefits of CI implementation

S. No	Benefits	Nominal gain (1)	Reasonable gain (2)	High gain (3)	Extremely high gain (4)
	Organization achievement				
1	Achieving high return on capital employed	1	2	3	4
2	Development of core-level competencies	1	2	3	4
3	Increased profit	1	2	3	4
4	Improved business performance	1	2	3	4
5	Enhanced professionalism in the organization	1	2	3	4
6	Improvement in competitive image of the organization	1	2	3	4
	Productivity				
7	Enhanced production rates	1	2	3	4
8	Improvement in overall equipment effectiveness	1	2	3	4
9	Reduced material and operator movement	1	2	3	4
10	Effective process flow	1	2	3	4
11	Reduced setup time	1	2	3	4
12	Reduced cycle time	1	2	3	4
	Cost				
13	Reduced operating cost	1	2	3	4
14	Reduced cost of consumables	1	2	3	4
15	Reduced cost of overhead expenditure	1	2	3	4
16	Reduced cost of inventories	1	2	3	4
17	Reduced inspection cost	1	2	3	4
	Quality				
18	Reduced customer complaints and rejections	1	2	3	4
19	Effective utilization of quality tools	1	2	3	4
20	Reduced scrap and rework	1	2	3	4
21	Reduced operator mistakes	1	2	3	4
22	Reduced total percentage defectives	1	2	3	4
23	Improved quality of work-life	1	2	3	4
	Delivery				
24	Improved delivery reliability	1	2	3	4
25	Providing faster delivery of the product	1	2	3	4

S. No	Benefits	Nominal gain (1)	Reasonable gain (2)	High gain (3)	Extremely high gain (4)
26	Customer satisfaction by providing at a time delivery	1	2	3	4
	Safety				
27	Reduced number of accidents	1	2	3	4
28	Maintaining health and safety standards	1	2	3	4
If any more, please specify					

Signature of the Respondent

Bibliography

Abdolshah, M., & Jahan, A. (2006). How to use continuous improvement tools in different life periods of organization. *International Conference on Management of Innovation and Technology 2*, 772–777, Singapore: IEEE.

Adams, M., Componation, P., Czarnecki, H., & Schroer, B. J. (1999). Simulation as a tool for continuous process improvement. *Proceedings of Winter Simulation Conference*, Phoenix, USA, 1, 766–773.

Agus, A., & Hassan, Z. (2011). Enhancing production through total quality management: Strategies for competitive advantage. *Procedia Social and Behavioral Sciences, 24*(1), 1650–1662.

Ahmed, S., Hassan, M. H., & Fen, Y. H. (2005). Performance measurement and evaluation in an innovative modern manufacturing system. *Journal of Applied Sciences, 5*(2), 385–401.

Ahuja, I. P. S., & Khamba, J. S. (2008). Total productive maintenance: Literature review and direction. *International Journal of Quality and Reliability Management, 25*(7), 709–756.

Al-Harbi, K. M. A. (2001). Application of AHP in project management. *International Journal of Project Management, 19*, 19–27.

Almeanazel, O. T. R. (2010). Total productive maintenance review and overall equipment effectiveness measurement. *Jordan Journal of Mechanical and Industrial Engineering, 4*(4), 517–522.

Ashmore, C. (2001). Kaizen–and the art of motorcycle manufacture. *Manufacturing Engineer, 80*(5), 220–222.

Augsdorfer, P., & Harding, R. (1995). Changing competitive forces in Europe: Continuous improvement in a sample of French, German and British companies. *European Business Review, 95*(4), 3–9.

Babbar, S. (2005). Service quality and business ethics. *International Journal of Services and Operations Management, 1*(3), 203–219.

Bahurmoz, A. (2006). The analytic hierarchy process at Dar Al-Hekmah, Saudi Arabia. *Interfaces, 33*(4), 70–78.

Bamber, C. J., Sharp, J. M., & Hides, M. (1999). Factors affecting successful implementation of total productive maintenance: A UK manufacturing case study perspective. *Journal of Quality in Maintenance Engineering, 5*(3), 162–181.

Bayazit, O., & Karpak, B. (2007). An analytical network process based framework for successful total quality management: An assessment of Turkish manufacturing industry readiness. *International Journal of Production Economics, 105*(1), 79–96.

Bayo-Moriones, A., Bello-Pintado, A., & Merino-Diaz de Cerio, J. (2010). 5S use in manufacturing plants: Contextual factors and impact on operating performance. *Journal of Quality and Reliability Management, 27*(2), 217–230.

J. Singh, H. Singh, *Strategic Implementation of Continuous Improvement Approach*, SpringerBriefs in Operations Management,
https://doi.org/10.1007/978-3-319-93121-0

Ben-Daya, M., & Raouf, A. (1996). A revised failure mode and effect analysis model. *International Journal of Quality and Reliability Management, 13*(1), 43–47.
Bendell, T. (2006). A review and comparison of six sigma and the lean organizations. *The TQM Magazine, 18*(3), 255–262.
Berger, A. (1997). Continuous improvement and Kaizen: Standardization and organizational designs. *Integrated Manufacturing Systems, 8*(2), 110–117.
Berggren, C. (1994). Nummi vs. Uddevalla. *Sloan Management Review* winter, 12–16.
Bessant, J. (2000).Creating and maintaining high involvement innovation. *Seminar on kaizen from understanding to action*, 30 March, Institute of Electrical Engineers, London.
Bessant, J. (2000). Developing and sustaining employee involving in continuous improvement. *IEE Seminar, Kaizen: from understanding to action (Ref. No. 2000/035)*, London, 2, pp. 1–18.
Bessant, J., & Caffyn, S. (1994). Rediscovering continuous improvement. *Technovation, 14*(1), 17–29.
Bessant, J., & Caffyn, S. (1997). High involvement innovation through continuous improvement. *International Journal of Technology Management, 14*(1), 7–28.
Bessant, J., Caffyn, S., & Gallagher, M. (2001). An evolutionary model of continuous improvement behavior. *Technovation, 21*(2), 67–77.
Bessant, J., & Francis, D. (1999). Developing continuous improvement capability. *International Journal of Operations and Production Management, 19*(11), 1106–1119.
Bhat, K. S., & Rajashekhar, J. (2009). An empirical study of barriers to TQM implementation in Indian Industry. *The TQM Magazine, 21*(3), 261–272.
Bhuiyan, N., & Baghel, A. (2005). An overview of continuous improvement: from the past to the present. *Management Decision, 43*(5), 761–771.
Bhuiyan, N., Baghel, A., & Wilson, J. (2006). A sustainable continuous improvement methodology at an aerospace company. *International Journal of Productivity and Performance Management, 55*(8), 671–687.
Black, S. A., & Porter, L. J. (1996). Identification of the critical factors of TQM. *Decision Sciences, 27*(1), 1–21.
Blanchard, B. S. (1997). An enhanced approach for implementing total productive maintenance in the manufacturing environment. *Journal of Quality in Maintenance Engineering, 3*(2), 69–80.
Boer, H., Berger, A., Chapman, R., & Gertsen, F. (Eds) (2000). *CI changes: From suggestion box to organisational learning. Continuous improvement in Europe and Australia*. Aldershot: Ashgate Publishing.
Boer, H., Caffyn, S., Chapman, R., Corso, M., & Coughlan, P. (1999). Continuous product innovation and knowledge management: the CIMA supporting methodology. *6th International product development management conference*, EIASM, Cambridge.
Bolton, M., & Heap, J. (2002). The myth of continuous improvement. *Work Study, 51*(6), 309–313.
Bon, A. T. B., & Karim, N. (2011). Total productive maintenance application to reduce defects of product. *Journal of Applied Sciences Research, 7*(1), 11–17.
Bond, T. C. (1999). The role of performance measurement in continuous improvement. *International Journal of Operations and Production Management, 19*(12), 1318–1334.
Bounds, G., Yorks, L., Adams, M., & Ranney, G. (1994). *Beyond total quality management toward the emerging paradigm*. New York, NY: McGraw-Hill.
Bradley, J., & Willett, J. (2004). Cornell students participate in lord corporation's Kaizen projects. *Interfaces, 34*(6), 451–459.
Brunet, A. P., & New, S. (2003). Kaizen in Japan: an empirical study. *International Journal of Operations and Production Management, 23*(12), 1426–1446.
Brunet, P. (2000). Kaizen in Japan. *IEE Seminar, Kaizen: from understanding to action (Ref. No. 2000/035)*. London, pp. 1–10.
Burns, A. (2000). Choosing the right tool from the tool box-some examples of Gemba Kaizen practice. *IEE Seminar, Kaizen: from understanding to action (Ref. No. 2000/035)*, London, 6, pp. 1–10.

Caffyn, S. (1999). Development of a continuous improvement self-assessment tool. *International Journal of Operations and Production Management, 19*(11), 38–53.
Carlson, J. G. (1989). JIT applications to warehousing operations. *Engineering Costs and Production Economics, 17*, 315–322.
Chakrborty, A., Bhattacharya, M., Ghosh, S., & Sarkar, G. (2013). Importance of KAIZEN concept in medium manufacturing enterprises. *International Journal of Management and Strategy, 4*(6), 234–245.
Chan, F. T. S., Lau, H. C. W., Ip, R. W. L., Chan, H. K., & Kong, S. (2005). Implementation of total productive maintenance: A case study. *International Journal of Production Economics, 95*(1), 71–94.
Chandrasekaran, M., Kannan, S., & Pandiaraj, P. (2008). Quality improvement in automobile assembly production line by using Kaizen. *Manufacturing Technology Today, 7*(3), 33–38.
Chang, R. (1995). Core threads of continuous improvement. *Management Development Review, 8*(4), 14–16.
Chapman, R. L., & Hyland, P. W. (1997). Continuous improvement strategies across selected manufacturing sector. *Benchmarking for Quality Management and Technology, 4*(3), 175–188.
Chase, R. B., & Stewart, D. M. (1994). Make your service fail-safe. *Sloan Management Review, 35*(3), 35–44.
Chaudhari, S. (1997). Kaizen at Morris electronics: Key to competitive success. *Portland international conference on management and technology*, Portland, p. 365.
Chen, C. I., & Wu, C. W. (2004). A new focus on overcoming the improvement failure. *Technovation, 24*, 585–591.
Chen, J. C., Dugger, J., & Hammer, B. (2000). A KAIZEN based approach for cellular manufacturing design: A case study. *The Journal of Technology Studies, 27*(2), 19–27.
Cheng, J. L. (2008). Implementing six sigma via TQM improvement: an empirical study of Taiwan. *The TQM Journal, 20*(3), 182–195.
Chengalvarayn, G., & Parker, S. C. (1991). Simulation analysis of just in time feasibility in a manufacturing environment. *Computer Industrial Engineering, 21*(14), 303–306.
Cheser, R. N. (1998). The effect of Japanese KAIZEN on employee motivation in US manufacturing. *International Journal Organizational Analysis, 6*(3), 197–212.
Chin, K., Wang, Y., Poon, G. K. K., & Yang, J. (2009). Failure mode effect analysis using a group-based evidential reasoning approach. *Computer and Operations Research, 36*(1), 1768–1779.
Cook, E., & Dale, B. G. (1995). Organization of continuous improvement: An examination. *The TQM Magazine, 7*(1), 7–13.
Davenport, T. H., & Short, J. E. (1990). The new industrial engineering: Information technology and business process redesign. *Sloan Management Review, 31*(4), 11–27.
Dehghan, R., Zohrab, M., Momeni, A., & Hoseini, M. (2006). Continuous improvement approach in the agriculture sector. *Asian Pacific Productivity Conference*, Seoul National University, South Korea.
Deming, W. E. (1995). *The new economics for industry government and education* (2nd ed.). Combridge, MA: MIT Press.
Desai, T. N., & Shrivastava, R. L. (2008). Six Sigma—A new direction to quality and productivity management. *Proceedings of the World Congress on Engineering and Computer Science*, San Fransisco, USA.
Deshmukh, S. V., & Chavan, A. (2012). Six Sigma and SMEs: A critical review of literature. *International Journal of Lean Six Sigma, 3*(2), 157–167.
Deshmukh, S. V., & Lakhe, R. R. (2010). Six Sigma awareness in Central Indian SMEs. *International Journal of Productivity and Quality Management, 5*(2), 200–212.
Dhongade, P. M., Singh, M., & Stouty, V. A. (2013). A review: Literature survey for the Implementation of Kaizen. *International Journal of Engineering and Innovative Technology, 3*(1), 57–60.
Dixon, N. (1994). *The organizational learning cycle: How we can learn collectively, developing organizations series*. London: McGraw-Hill.

Dogra, M., Sharma, V. S., Sachdeva, A., & Dureja, J. S. (2011). TPM-A key strategy for productivity improvement in process industry. *Journal of Engineering Science and Technology, 6*(1), 1–16.

Doolen, T. L., June, W. Q., Akan, V., Eileen, M., & Jennifer, F. (2003). Development of an assessment approach for KAIZEN Events. *Proceedings of the 2003 Industrial Engineering and Research Conference*, CD-ROM.

Dulhai, G. (2008). The 5S strategy for continuous improvement of the manufacturing processes in Autocar Exhaust. *Management and Marketing, 3*(4), 115–120.

Dwyer, J. (2006). Seeing is believing. *Works Management, 59*(10), 54–57.

Elmuti, D., & Kathawala, Y. (1997). An overview of benchmarking process: A tool for continuous improvement and competitive advantage. *Benchmarking for Quality Management and Technology, 4*(4), 229–243.

Erlandson, R. F., Noblett, M. J., & Phelps, J. A. (1998). Impact of poka-yoke device on job performance of individuals with cognitive impairments. *IEEE Transactions on Rehabilitation Engineering, 6*(3), 269–276.

Farris, J. A., VanAken, E. M., Doolen, T. L., & June, W. (2009). Critical success factor human resource outcome in Kaizen event: An Empirical study. *International Journal of Production Economics, 117*(1), 42–65.

Gallagher, M., Austin, S., & Caffyn, S. (1997). *Continuous improvement in action: The journey of eight companies*. London: Kogan Page.

Gapp, R., Fisher, R., & Kobayash, K. (2008). Implementing 5S within a Japanese context: an integrated management system. *Management Decisions, 46*(4), 565–579.

Garcia-Lorenzo, A., Prado, J., & Arca, J. G. (2000). Continuous improvement and employee participation in SMEs. *The TQM Magazine, 12*(4), 290–294.

Ghalayani, A. M., Noble, J. S., & Crowe, T. J. (1997). An integrated dynamic performance measurement system for improving manufacturing competitiveness. *International Journal of Production Economics, 48*(2), 20–25.

Gibb, A., & Davies, L. (1990). In pursuit of frameworks for the development of growth models of the small business. *International Small Business Journal, 9*(1), 15–31.

Gonsalves, G. C. (2002). Business process management: Integration of quality management and reengineering for enhanced competitiveness. *Pro Quest Information and Learning Company, 7*(1), 120–128.

Goriwondo, W. M., Mhlanga, S., & Kazembe, T. (2011). Optimizing a production system using tools of total productive maintenance: Datlabs pharmaceuticals as a case study. *Proceedings of the International Conference on Industrial Engineering and Operations Management*, Kuala Lumpur, Malaysia.

Granja, D. A., Picchi, F. A., & Robert, G. T. (2005). Target and Kaizen costing in construction. *Proceedings IGLC-13*, Sydney, Australia, pp. 227–233.

Grunberg, T. (2003). A review of improvement methods in manufacturing operations. *Work Study, 52*(2), 89–93.

Gunasekaran, A., Korukonda, A. R., Virtanen, I., & Yli-Olli, P. (1994). Improving productivity and quality in manufacturing organizations. *International Journal of Production Economics, 36*(2), 169–183.

Hackman, J. R., & Oldham, G. R. (1980). *Work redesign*. Reading, MA: Addison-Wesley.

Hamel, G., & Prahalad, C. K. (1994). *Competing for the future*. Boston, MA: Harvard Business School Press.

Hammer, M., Champy, J., & Tathan, R. L. (1993). *Reengineering the corporation: A manifesto for business revolution*. New York: Harper Collins.

Harrington, H. J. (1995). Continuous versus breakthrough Improvement: finding the right answer. *Business Process Reengineering and Management Journal, 1*(3), 31–49.

Hongming, H., Sun, H., & Xu, Y. (2000). An empirical study on quality management practices in Shinghai manufacturing industries. *Total Quality Management, 11*(8), 1111–1122.

Huang, P. Y., Rees, P. L., & Taylor, B. W. (1983). A simulation analysis of Japanese JIT for a multimachine, multistage production system. *Decision Science, 14*(2), 326–343.

Hyland, P., Mellor, R., Omara, E., & Kondepndi, R. (2000). A comparison of Australian firms and their use of continuous improvement tools. *The TQM Magazine, 12*(2), 117–124.

Imai, M. (1986). *KAIZEN: The key to Japan's competitive success*. New York: McGraw Hill.

Imai, M. (1997). *Gemba Kaizen: A commonsense, low cost approach to management*. New York: McGraw Hill.

Ionica, A. C., Leba, M., & Corbu, C. (2013). Professional Development—Between Reengineering and Continuous Improvement. *2nd World Conference on Educational Technology Researches—WCETR2012*.

Irane, Z., & Sharp, J. M. (1997). Integrating continuous improvement and innovation into a corporate culture: a case study. *Technovation, 17*(4), 225–226.

Jaca, C., Viles, E., Mateo, R., & Santos, J. (2012). Components of sustainable improvement systems: theory and practices. *The TQM Journal, 24*(2), 142–154.

Jager, B., Minnie, C., Jager, J., Welgemoed, M., Bessant, J., & Francis, D. (2004). Enabling continuous improvement: a case study of implementation. *Journal of Manufacturing Technology Management, 15*(4), 315–324.

Jayaraman, A., Green, J. A., & Gunal, A. K. (1995). Continuous improvement applied to simulation modeling: A case study. *Proceedings of winter simulation conference,* Arlington, USA, pp. 930–935.

Jha, S., Michela, J., & Noori, H. (1999). Within the Canadian boundaries: A close look at Canadian Industries implementing continuous improvement. *The TQM Magazine, 11*(3), 188–197.

Jha, S., Noori, H., & Michela, J. (1996). The dynamics of continuous improvement. *International Journal of Quality Science, 1*(1), 19–47.

Jonsdottir, S., Ingason, H. I., & Jonasson, H. I. (2014). Continuous improvement projects in certified organizations in Iceland: traditional projects or not*? Social and Behavioral Sciences, 119*, 142–151.

Jorgensen, F., Boer, H., & Gertsen, F. (2003). Jump-starting continuous improvement through self-assessment. *International Journal of Operations and Production Management, 23*(10), 1260–1278.

Jung, J. Y., Wang, Y. J., & Wu, S. (2009). Competitive strategy, TQM practice, and Continuous Improvement of international project management. *International Journal of Quality and Reliability Management, 26*(2), 164–183.

Kanji, G. K., & Asher, M. (1996). *100 Methods for total quality management*. London: Sage Publications.

Karkoszka, T., & Szewieczek, D. (2007). Risk of the processes in the aspect of quality, natural environment and occupational safety. *Journal of Achievements in Materials and Manufacturing Engineering, 20*(3), 539–542.

Kaskhedikar, K., & Katore, H. V. (2013). Improvement of quality and productivity in casting industries. *Mechanica Confab,* 2(3), 63–71.

Kaushik, P. (2012). Relevance of six sigma line of attack in SMEs: A case study of a die casting manufacturing unit. *Journal of Engineering and Technology, 1*(2), 107–112.

Kaye, M., & Anderson, R. (1999). Continuous improvement: The ten essential criteria. *International Journal of Quality and Reliability Management, 16*(5), 485–506.

Kaye, M. M., & Dyason, M. D. (1995). The fifth era. *The TQM Magazine, 7*(1), 33–37.

Kelle, P., & Schneider, H. (1992). Extension of a reliability-type inventory model to just in time systems. *International Journal of Production Economics, 26*(1), 319–326.

Ketokivi, M. A., & Schroeder, R. G. (2004). Perceptual measures of performance: Fact or fiction. *Journal of Operations Management, 22*, 247–262.

Khan, I. A. (2011). KAIZEN: The Japanese strategy for continuous improvement. *VSRD International Journal of Business and Management Research, 1*(3), 177–184.

Khanna, V. K. (2009). 5S and TQM status in Indian organization. *The TQM Journal, 21*(5), 486–501.

Kikuchi, K., Kikuchi, T., & Takai, T. (2007). Method of overall consumable effectiveness. *IEEE International symposium on semiconductor manufacturing*, Santa Clara, USA, pp. 1-4.

Kim, W. C., & Mauborgne, R. (1999). Strategy, value innovation and the knowledge economy. *Sloan Management Review*, Spring, 41–54.

Koskela, L. (1999). Management of production in construction: a theoretical view. *Proceedings of seventh annual conference of the International Group for lean Construction*, University of California, CA, pp. 241–252.

Kossoff, L. L. (1993). Total quality or total chaos? *HR Magazine, 38*(4), 131–134.

Lawler, E. E. (1991). *High involvement management*. San Francisco, CA: Jossey-Bass.

Lawrence, J. L. (1999). Use mathematical modeling to give your TPM implementation effort an extra boost. *Journal of Quality in Maintenance Engineering, 5*(1), 62–69.

Lee, M. (2000). Customer service excellence through people motivation and Kaizen. *IEE Seminar, Kaizen: from understanding to action (Ref. No. 2000/035)*, 5, pp. 1–21.

Lee, S. S., Dugger, J. C., & Chen, J. C. (2000). Kaizen: An essential tool for inclusion in industrial technology curricula. *Journal of Industrial Technology, 16*(1), 1–7.

Leede, J. D., & Looise, J. K. (1999). Continuous improvement and the mini-company concept. *International Journal of Operations and Production Management, 19*(11), 1188–1202.

Leem, C. S., & Kim, J. (2004). An integrated evaluation system based on the continuous improvement model of IS performance. *Industrial Management and Data System, 104*(2), 115–128.

Lillrank, P. (1995). The transfer of management innovations from Japan. *Organization Studies, 16*(6), 971–989.

Lillrank, P., & Kano, N. (1989). *Continuous Improvement—quality control circles in Japanese Industry*. Ann Arbor, MI: University of Michigan.

Lixia, R. E., & Bo, M. (2008). Implementing 5S to Chinese Enterprise. *Journal of Operations Management, 24*(5), 29–47.

Ljungstrom, M. (2005). A mode for starting up and implementing continuous improvement and work development in practice. *The TQM Magazine, 17*(5), 385–405.

Locke, E. A., & Jain, V. K. (1995). Organizational learning and continuous improvement. *The International Journal of Organizational Analysis, 3*(1), 45–68.

Love, P. E. D., Gunasekaran, A., & Li, H. (1999). Improving the competitiveness of manufacturing companies by continuous incremental change. *The TQM Magazine, 10*(3), 177–185.

Macbeth, D. K. (1987). Supplier management in support of JIT activity: A research agenda. *International Journal of Operations & Production Management, 7*(4), 53–63.

Mackle, K. (2000). A frame work for implementation of Kaizen management system audit. *IEE Seminar, Kaizen: from understanding to action (Ref. No. 2000/035)*, 3, pp.1-6.

Malik, S. A., Li-bin, L., YeZhuang, T., & Xiao-lin, S. (2007) Continuous improvement practices in Asian developing countries: A comparative analysis between Chinese and Pakistani manufacturing industry. *14th International conference on management science and engineering*, Harbin, P.R. China, pp. 692–697.

Malik, S. A., & YeZhuang, T. (2006). Execution of continuous improvement practices in Spanish and Pakistani industry: A comparative analysis. *IEEE International Conference on Management of Innovation and Technology, Singapore*, 2, pp. 761–765.

Mandal, A., & Deshmukh, S. G. (1994). Vendor selection using interpretive structural modelling (ISM). *International Journal of Operations & Production Management, 14*(6), 52–59.

Marin-Garcia, J. A., Val, M. P., & Martin, T. B. (2008). Longitudinal study of the results of continuous improvement in an industrial company. *Team Performance Management, 14*(1/2), 56–69.

Martin, D. (1993). Toward Kaizen: The quest for quality improvement. *Library Management, 14*(4), 4–12.

Matsui, Y. (2007). An empirical analysis of just in time production in Japanese manufacturing companies. *International Journal of Production Economics, 108*(1), 153–164.

McAdam, R., Stevenson, P., & Grem, A. (2000). Innovation change management in SMEs: Beyond continuous improvement. *Logistics Information Management, 13*(3), 138–149.

Michalska, J., & Szewieczek, D. (2007). The 5S methodology as a tool for improving the organiosation. *Journal of Achievements in Materials and Manufacturing Engineering, 24*(2), 211–214.

Monden, Y. (1998). *Toyota production system: An integrated approach to just-in-time* (3rd ed.). Routledge, London: Industrial Engineering and Management Press.

Montabon, F. (2005). Using Kaizen events for back office processes: The recruitment of frontline supervisor co-ops. *Total Quality Management & Business Excellence, 16*(10), 1139–1147.

Mora, E. (2002). *The right ingredients for a successful TPM or lean implementation* (2nd ed.). New York, NY: Industrial Press.

Motwani, J. (2001). Critical factors and performance measures of TQM. *The TQM Magazine, 13*(4), 292–300.

Moubray, J. (1997). *Reliability centered maintenance* (2nd ed.). New York, NY: Industrial Press.

Nakajima, S. (1988). *Introduction to TPM*. Cambridge, MA: Productivity Press.

Natarajan, M., Senthil, V., Devadasan, S. R., & Mohan, N. V. (2011). QUARNEWSS: A model for applying six sigma framework to achieve continuous quality and reliability improvement in new product development. *International Journal of Productivity and Quality Management, 7*(1), 44–73.

Newitt, D. J. H. (1996). Beyond BPR and TQM- managing the processes: Is Kaizen enough? In *Proceedings Industrial Engineering*, London, Institution of Electric Engineers, pp. 1–5.

Oakley, J. (1996). Communicating for competitive advantage: A continuous improvement model. *Journal of Communication Management, 1*(2), 169–173.

Ohno, T. (1988). *Toyota production system: Beyond large-scale production*. Cambridge, MA: Productivity Press.

Oliver, N. (1990). The dynamic of just in time. *New Technology Work and Employment, 3*(2), 19–27.

Oprime, P. C., Mendes, G. H., & Pimenta, M. L. (2012). Continuous improvement: Critical factors in Brazilian industrial companies. *International Journal of Productivity and Performance Management, 61*(1), 69–92.

Osada, T. (1991). *The 5-S: Five keys to a total quality environment*. Tokyo: Asian Productivity Organization.

Palmer, V. S. (2001). Inventory management Kaizen. *Proceedings 2nd International Workshop on Engineering Management for Applied Technology*, Austin, pp. 55–56.

Park, K. S., & Han, S. W. (2001). TPM—total productive maintenance: Impact on competitiveness and a framework for successful implementation. *Human Factors and Ergonomics in Manufacturing, 11*(4), 321–338.

Pranckevicius, D., Diaz, D. M., & Gitlow, H. (2008). A lean six sigma case study: An application of the '5S' technique. *Journal of Advances in Management Research, 5*(1), 63–79.

Radharamanan, R., Godoy, L. P., & Watanabe, K. I. (1996). Quality and productivity improvement in a custom-made furniture industry using Kaizen. *Computer and Industrial Engineering, 31*(1/2), 471–474.

Ramayan, T., Jantan, M., & Hassan, M. M. (2002). Change management and implementation of total productive maintenance: An exploratory study of Malaysian manufacturing companies. *Change Management, 3*(1), 35–49.

Reid, R. A. (2006). Productivity and quality improvement: An implementation framework. *International Journal of Productivity and Quality Management, 1*(1/2), 26–36.

Robinson, A. G. (1991). *continuous improvement in operations —a systematic approach to waste reduction*. Cambridge, MA: Productivity Press.

Rusiniak, S. (1996). Maximizing your IE value. *Industrial Engineer, 28*(6), 12–16.

Sabhu, R. U., & Dalu, R. S. (2011). Six Sigma implementation in Indian medium scale automotive enterprises—a review and agenda for future research. *International Journal of Six Sigma and Competitive Advantage, 6*(3), 224–247.

Salah, S., Rahim, A., & Carretero, J. A. (2010). The integration of six sigma and lean management. *International Journal of Lean Six Sigma, 1*(3), 249–274.

Saleem, M., Khan, N., Hameed, S., & Ch, M. A. (2012). An analysis of relationship between total quality management and TQM. *Life Science Journal, 9*(3), 31–40.

Sankar, N. R., & Prabhu, B. S. (2001). Modified approach for prioritization of failures in a system failure mode and effects analysis. *International Journal of Quality & Reliability Management, 18*(3), 324–335.

Santos, A., Powell, J. A., & Formoso, C. T. (2000). Setting stretch targets for driving continuous improvement in construction: analysis of Brazilian Company. *Work Study, 49*(2), 50–58.

Savino, M. M., & Mazza, A. (2014). Toward environmental and quality sustainability-an integrated approach for continuous improvement. *IEEE Transactions on Engineering Management, 61*, 28–34.

Savolainen, T. I. (1999). Cycles of continuous improvement: Realizing competitive advantage through quality. *International Journal of Operation and Production Management, 19*(11), 1203–1222.

Schroeder, R. G. (1993). *Operations management: Decision making in the operations function* (4th ed.). New York, NY: McGraw-Hill.

Seaker, R., & Waller, M. A. (1996). The common thread in TQM, empowerment, reengineering and continuous improvement. *International Journal of Quality and Reliability Management, 13*(1), 24–31.

Setijono, D. (2010). Model and principles of stakeholders-oriented quality management based on radical (discontinuous) improvement—a modern re-interpretation of TQM and CWQC? *International Journal of Quality and Innovation, 1*(1), 167–183.

Shaari, J. A. N. (2010). Barriers to implement TQM in Japanese way: a study on companies in Malaysia. *International Review of Business Research Papers, 6*(5), 400–410.

Sharma, R. K., Kumar, D., & Kumar, P. (2006). Manufacturing excellence through TPM implementation: a practical analysis. *Industrial Management & Data System, 106*(2), 256–280.

Sheridan, J. H. (1997). Kaizen Blitz. *Industry Week, 246*(16), 18–27.

Shingeo, S. (1988). *Non-stock production: The shingo system for continuous improvement.* Cambridge, MA: Productivity Press.

Shingo, S. (1985). *A Revolution in manufacturing: The SMED System.* Andrew Oillon: Productivity Press.

Siekman, P. (2002). Pentair fixes its own mess. *Fortune, 146*(6), 156–160.

Sim, K. L., & Rogers, J. W. (2009). Implementing lean production systems: barriers to change. *Management Research News, 32*(1), 37–49.

Singh, B., & Khanduja, D. (2010). SMED for quick changeover in foundry SMEs. *International Journal of Productivity and Performance Management, 59*(1), 98–116.

Singh, H., & Khamba, J. S. (2010). An empirical examination for enhancing the utilization level of advanced manufacturing technologies in India. *Journal of Advances in Management Research, 7*(1), 112–126.

Soderquist, K. (1996). Managing innovation in SMES: A comparison of companies in UK, France and Portugal. *International Journal of Technology Management, 12*(3), 291–305.

Sohal, A. S., & Terzioviski, M. (2000). TQM in Australian manufacturing: factors critical to success. *International Journal of Quality and Reliability Management, 17*(2), 158–167.

Sokovic, M., Pavletic, D. & Fakin, S. (2005). Application of Six Sigma methodology for process design quality improvement. *13th International scientific conference on achievements in mechanical and material engineering,* Poland, pp. 611–614.

Soti, A., Kaushal, O. P., & Shankar, R. (2011). Modeling the barriers of six sigma using interpretive structural modeling. *International Journal of Business Excellence, 4*(1), 94–110.

Starkey, K., & Mackinlay, A. (1993). *Ford and the search for competitive advantage, strategy and the human resource*. Oxford: Blackwell.

Sushil, X. (1999). *Flexibility in management, global institute of flexible systems management,* New Delhi.

Suzaki, K. (1987). *The new manufacturing challenge-Techniques of manufacturing systems.* New York: John Wiley and Sons, Inc..

Tahla, M. (2004). TQM-An overview. *The Bottom Line: Managing Library Finances, 17*(1), 15–19.

Tanik, M. (2010). Improving 'order handling' process by using QFD and FMEA methodologies: a case study. *International Journal of Quality and Reliability Management, 27*(4), 404–423.

Teian, K. (1992). *Guiding continuous improvement through employee suggestions*. Portland, OR: Productivity Press.

Teng, S. G., Ho, S. M., Shumal, D., & Liu, P. C. (2006). Implementing FMEA in a collaborating supply chain environment. *International Journal of Quality and Reliability Management, 23*(2), 179–196.

Terziovski, M. (2001). The effect of continuous improvement and innovation management practices on small to medium performance. *Proceedings of 5th International Conference on Quality and Innovation Management*, Euro-Australian Co-operation Centre for Global Management, pp. 1–22.

Terziovski, M. (2002). Achieving performance excellence through an integrated strategy of radical innovation and continuous improvement. *Measuring Business Excellence, 6*(2), 5–14.

Terziovski, M., & Power, D. (2007). Increasing ISO 9000 certification benefits: A continuous improvement approach. *International Journal of Quality and Reliability Management, 24*(2), 141–163.

Terziovski, M., & Sohal, A. S. (2000). The adoption of continuous improvement and innovation strategies in Australian manufacturing firms. *Technovation, 20*(10), 539–550.

Thakkar, J., Kanda, A., & Deshmukh, S. G. (2008). Interpretive structural modeling (ISM) of IT-enablers for Indian manufacturing SMEs. *Information Management & Computer Security, 16*(2), 113–136.

Tomlinson, Y. (1994). Involving staff in the process of continuous improvement. *Training for Quality, 2*(3), 22–25.

Tonnessen, T. (2005). Continuous innovation through wide employee participation. *The TQM Magazine, 17*(2), 195–207.

Tsang, J. H. Y., & Antony, J. (2001). Total quality management in UK service organizations: Some key findings from a survey. *Managing Service Quality, 11*(2), 132–141.

Tseng, M. L., Chiu, A. S. F., & Chinag, J. H. (2006). The relationship of continuous improvement and cleaner production on operational performance: An empirical study in electronic manufacturing firms, Taiwan, China. *International Journal of Management Science and Engineering Management, 1*(1), 71–80.

Van Der Wal, R. W. E., & Lynne, D. (2002). Total productive maintenance in a South African pulp and paper company: A case study. *The TQM Magazine, 14*(6), 359–366.

Vassie, L. (1998). A proactive team-based approach in health and safety management. *Employee Relations, 20*(6), 577–593.

Walsh, A., Hughes, H., & Maddox, D. P. (2002). Total quality management continuous improvement: Is the philosophy a reality. *Journal of European Industrial Training, 26*(6), 299–307.

Warfield, J. W. (1974). Developing interconnected matrices in structural modelling. *IEEE Transcript on Systems, Men and Cybernetics, 4*(1), 51–81.

Watson, M. (1986). *The deming management method.* New York, NY: Perigee Books.

Watson, R. M. (1998). Implementing self managed process improving team in continuous improvement environment. *The TQM Magazine, 10*(4), 246–257.

Wemmerlov, U., & Hyer, N. (1989). Cellular manufacturing in US industry: A survey of users. *International Journal of Production Research, 27*(9), 1511.

Werner, K. (1994). Act and think to create the learning organization of the 1990's. *European Management Journal, 12*(3), 11–31.

Wickens, P. D. (1990). Production management: Japanese and British approaches. *IEE Proceedings A, Science, Measurement and Technology, 137*(1), 52–54.

Williams, M. (2001). Maximum cost reduction minimum effort. *Manufacturing Engineer, 80*(4), 179–182.

Williamson, A. (1997). Target and Kaizen costing. *Manufacturing Engineer, 76*(1), 22–24.

Wittenberg, G. (1994). Kaizen-The many ways of getting better. *Assembly Automation, 14*(4), 12–17.
Womack, J. P., & Jones, D. T. (1996). *Lean thinking*. New York: Simon &Schuster.
Woods, J. (1997). The six values of a quality culture. *National Productivity Review, 16*(2), 49–54.
Yamashina, H. (1995). Japanese manufacturing strategy and the role of total productive maintenance. *Journal of Quality in Maintenance Engineering, 1*(1), 27–38.
Yang, C., Lin, W., Lin, M., & Huang, J. (2006). A study on applying FMEA to improving ERP introduction: An example of semiconductor related industries in Taiwan. *International Journal of Quality and Reliability Management, 23*(3), 298–322.
Yang, J., Huang, H., He, L., Zhu, S., & Wen, D. (2011). Risk evaluation in failure mode and effect analysis of aircraft turbine rotor blades using Dempster—shafer evidence theory under uncertainty. *Engineering Failure Analysis, 18*(1), 2084–2092.
Yan-jiang, C., Dan, W., & Lang, X. (2006). Influencing factors of continuous improvement and tendency to change. *IEEE International conference on management of innovation and technology,* vol. 1, Singapore, PP. 181–185.
Yan-jiang, C., Lang, X. & Xiao-na, W. (2006). Empirical study of influencing factors of continuous improvement. *International conference on management science and engineering*, Lille, France, PP. 577–581.
Yeo, C. H., Goh, T. N., & Xia, M. (1995). A positive management orientation for continuous improvement. *IEEE Conference on Engineering Management*, pp. 208–213.
Yoo, S. H., Kim, D., & Park, M. (2012). Inventory models for imperfect production and inspection processes with various inspection options under one-time and continuous improvement investment. *Computer and Operations Research, 39*(2), 2001–2015.
Yung, W. K. C. (1996). An integrated model for manufacturing process improvement. *Journal of Materials Processing Technology, 61*(1), 39–43.
Zeleny, M. (1982). *Multiple criteria decision making*. New York: McGraw Hill.
Zhang, Z., & Chu, X. (2011). Risk prioritization in failure mode effect analysis under uncertainty. *Expert Systems and Applications,., 38*(1), 206–214.

MIX
Papier aus verantwortungsvollen Quellen
Paper from responsible sources
FSC® C105338

If you have any concerns about our products,
you can contact us on
ProductSafety@springernature.com

In case Publisher is established outside the EU,
the EU authorized representative is:
Springer Nature Customer Service Center GmbH
Europaplatz 3, 69115 Heidelberg, Germany

Printed by Libri Plureos GmbH
in Hamburg, Germany